ENGINEERS IN BRITAIN

ENGINEERS IN BRITAIN

A Sociological Study of the Engineering Dimension

IAN A. GLOVER

MICHAEL P. KELLY

Dundee College of Technology
and
The University of Glasgow

London
ALLEN & UNWIN
Boston   Sydney   Wellington

Allen & Unwin, the academic imprint of
Unwin Hyman Ltd
PO Box 18, Park Lane, Hemel Hempstead, Herts HP2 4TE, UK
40 Museum Street, London WC1A 1LU, UK
37/39 Queen Elizabeth Street, London SE1 2QB

Allen & Unwin Inc.,
8 Winchester Place, Winchester, Mass. 01890, USA

Allen & Unwin (Australia) Ltd,
8 Napier Street, North Sydney, NSW 2060, Australia

Allen & Unwin (New Zealand) Ltd in association with
the Port Nicholson Press Ltd,
60 Cambridge Terrace, Wellington, New Zealand

First published in 1987

British Library Cataloguing in Publication Data

Glover, Ian A.
  Engineers in Britain: a sociological study of the engineering dimension.
1. Engineering firms — Great Britain — Employees
2. Industrial sociology — Great Britain
I. Title II. Kelly, Michael P.
305.9′62′0941   TA157
ISBN 0-04-301222-1
ISBN 0-04-301223-X Pbk

Library of Congress Cataloging in Publication Data

Glover, Ian.
  Engineers in Britain
Bibliography: p.
Includes indexes.
1. Engineers — Great Britain. 2. Engineering —
Social aspects — Great Britain. I. Kelly, Michael P.
II. Title.
TA157.G57 1987   620′.00941   86-32284
ISBN 0-04-301222-1 (alk. paper)
ISBN 0-04-301223-X (pbk.: alk. paper)

Typeset in 10 on 11 point Bembo by Fotographics (Bedford) Ltd
and printed in Great Britain by Billing & Sons Ltd, London and Worcester

To Our Parents

Contents

List of Tables and Figures

Tables

Figures

Preface and Acknowledgements

This book represents the combined efforts of two sociologists to wrestle with the problem of introducing sociological ideas to non-sociologists. Our particular chosen audience consists of practitioners and students of engineering but we hope that the book will also appeal to all those with an interest in engineers and engineering too. We are bold enough to hope that all thinking people are capable of realizing that they have such an interest.

Our problem in writing this book has revolved around the manner of presentation. Many introductory sociology books consist of a sort of Cook's tour or overview of major sociological writers on major sociological themes. There were two reasons why we felt that this would be an unsatisfactory format for this book. First, there are a great many other books in which this is generally accomplished better than it could be by us. Secondly, and more importantly, the overview format can sometimes lead to over-generality in approach and to a lack of sympathy on the part of the audience. At its worst such an approach can be rather like that of an optician who spends all the time describing and discussing lenses without ever using them to look at the world. We know this through long and sometimes tough experience of teaching sociology to a variety of non-sociologists.

We have adopted a different kind of framework in that we have tried to write a book which, while introducing both key sociological ideas and writers, attempts to demonstrate the importance of the ideas by applying them directly to specific problems. What we do is thus to apply sociological *reasoning*, more than themes or ideas but not excluding them as will be apparent, to some of the main social problems confronting individual engineers and engineering as an occupation.

The key issue in our book, and it is one derived from the sociological study of social institutions, is this. Engineering plays a central and crucial role in the productive and reproductive life of a society, particularly in that of an industrial society. However, in Britain, the first industrial country, although the productive role is crucial, engineers as individuals, the occupation itself, teachers of engineering, as well as academics writing about it, are often marginalized in terms of British economic, political and academic life. That is, their concerns have been deemed to be secondary, subordinate, derivative, unimportant, uninteresting. This book is concerned with this central dilemma or contradiction.

The problem which this leaves us with is that we are in effect trying to do two things simultaneously. We are trying to introduce ideas and to use them to produce higher-order explanations. We have tried to maintain the somewhat precarious balance between the two things although, having just completed the book, we remain uncertain as to whether we have fully succeeded. Even if our enterprise does not quite come off, we think that the issues we highlight are so important to the future of Britain, and possibly to other industrial societies which may be attracted to British-style pretensions, that our attempt is nevertheless valuable.

Many people have helped us to write this book. First and foremost are our students, mainly those whom we have taught who were studying engineering and business subjects in the public sector of British higher education. These have been the first audience for many of our ideas, with whom we worked many of them out in lectures, seminars and informal discussions. Critical response from this most demanding group of scrutineers has been a source of inspiration and desperation. Second, our families and friends, whose lives have been periodically disrupted by the heavy demands of our attempts to do serious research and writing, and who have still been supportive. Third, we thank our publishers Allen & Unwin and our agent Frances Kelly who have shown amazing patience in the face of repeated missed deadlines and improbable excuses as to why those deadlines had not been met. Fourth, there are the many individuals and colleagues who gave up their own time to read all or parts of our manuscripts, who offered advice and provided encouragement, or whose help was more implicit but nonetheless valuable and welcome. Two in particular deserve special mention. First there was Michael Fores, the long-distance iconoclast who has quietly inspired us and several better-known social scientists over the last fifteen years through his influential work on behalf of Britain's engineers on the one hand and with his witty common sense on the other. He has extensively read, criticized and breathed upon our drafts, adding (we believe) depth and point in several places and encouraging us to subtract facile or mistaken material in others. Second there was Pat Sawers, who typed the often none-too-legible final draft with her usual considerable speed and exemplary accuracy. We have had material typed by many people over the years in more than a few settings and Pat is the most helpful and competent we have known.

The others whom we thank, and who like the previous two individuals are not in any way guilty by their association with us are, in alphabetical order: Charles Abraham, Lord (John) Baker, Greg Bamber, Richard Bines, Bob Brewer, the late Lord (Wilfred) Brown, Geoff Beuret, Lynda, Sandra, Dorothy and Robert Cairns, Dennis Carnegie, John Child, Richard Crowder, Martin Dowling, John Downes, Helen Foord, Douglas Garbutt, Hector Gow, Ken Green, Clifford Healey, Peter Herriot, Stewart Howe (partly for his help and forbearance as our Head of Department, partly in his other capacity as Chairman of our College's Research Committee with its valued support), Liam Hudson, Roy Hutchison, Stanley Hutton, Ian Jones, Tessa Kelly, Peter Lawrence, John Levy, Keith Lockyer, Jim McArthur, the late Alan McConkey, Kevin McCormick, Alistair Mant, Graeme Martin, Ian Massie, John Mather, David May, Barbara Paterson, Corinne Peddie, Celia Pillay, Geoff Price, the late Gerry Reynolds, Pat Scrutton, Arndt Sorge, Sìne Stewart, K. W. Taylor, Stephen Wearne, David Weir, Peter Whalley and Denys Wood. A special kind of thanks is also due to the staff of our College's Library, especially Jenny Park, Shona Wood and Val McKay. Finally and in minor partial defence of the missed deadlines mentioned earlier, some of them were due to the rapid changes in the education and training of and other factors affecting Britain's engineers in the last few years. We mention this, however, more to illustrate what we feel to be the importance of our subject than as an excuse for ourselves.

September 1986

IAN GLOVER
MICHAEL KELLY

PART ONE

1 Introduction: Sociology and Engineering

Although the reasons are very different in each case, neither engineers nor sociologists have received much appreciation and recognition in Britain in the last third of the twentieth century. Nor do they appear to have had much respect for each other. Many engineers feel that sociology is politically motivated and unscientific; while sociologists impute blame to 'technologists' for many of the evils of industrialization such as the destruction of craft skills, exploitation, conspicuous waste, pollution, high levels of military spending, and for being servants of the irresponsibly powerful. This book is written in the belief that the majority of the criticisms, from both sides, are unfair or wrong, and in the hope that practitioners and teachers of engineering will come to see that the study of sociology has something to offer their students.

Sociology ought to be useful to engineers. A popular American definition of an engineer is 'someone who can make for one dollar what any bloody fool can make for two'. This suggests that engineers should be concerned with the value of their output and not only with whether their artefacts and machines work. Further, in the Finniston report on the engineering profession and its problems it was argued that engineers would increasingly need to equip themselves with knowledge of economic and 'social and political considerations and constraints' (Finniston, 1980). As a subject which explicitly sets out to help people to understand such things, sociology should be of direct use to the engineer. Indeed we would argue that there is little value in a sociology that does not help people to understand their world better.

The particular world we wish to help people to understand is the world of the engineer in Britain. We suggest that the relatively low status of British engineers and their work is one of the most important and interesting features of contemporary British society compared with other industrial ones. It is a sociological analysis of this issue and the reasons for it which constitute the major focus of this book.

The kind of sociology which we use for our analysis follows Berger and Kellner (1982). We are interested in social institutions as they are, not as we would like them to be. Our attitudes towards general theories of society are eclectic. We see elements both of continuity and change, and of consensus and conflict in the social world, and we regard people as being both makers and victims of history. We find the idea that there are final answers to questions about human behaviour dangerous and naive. We have more respect for sociological arguments which are informed by evidence rather than for self-perpetuating scholastic debates. In short we believe that sociology must first and foremost be concerned with society, not only with the study of ideas, and that sociological theories themselves can only develop properly if they are testable and grounded in real life events (Glaser and Strauss, 1968).

The sense in which we use the term sociology is as follows. At the most general level we, as sociologists, are interested in the study of people living and working in human societies. This does not nor is it intended to mean that sociology is some kind of superior mixture of history, politics, economics, geography, anthropology and psychology (although in our experience some professional socio-logists do seem to hold this view and are rightly ridiculed by other academics for doing so) but rather that as sociologists we are concerned with four principal questions. What is society? That is to say what is the distinctive nature of the collectivities, large and small, which human beings inhabit? What is the nature of the relationships between the various parts or elements in the collectivities such as individuals, groups, families and institutions? How, and perhaps more ambitiously why, do societies change? And how, given the apparently never-ending flux of human behaviour, is continuity and order possible in social relationships? (Useful introductions to sociology and sociological ideas, where various aspects of these questions are explored in detail, can be found in the following texts: Broom and Selznick, 1977; Cotgrove, 1978; Haralambos, 1980; and Hurd, 1986.)

Yet stating a series of general questions is still not sufficient for a full appreciation of what sociology is about. The sociologist assumes that there are certain regularities and patterns in human conduct and that these tend to repeat themselves. Therefore sociology articulates and analyses sometimes unsuspected continuities and regularities in human affairs. In other words, while everyday experience may not support this view, the world of human behaviour does not simply consist of random events: human behaviour has both purpose and structure.

The assumption of purpose and structure (which in fact is rather more than an assumption, since it is documented with evidence collected by sociologists and others over the last 200 years or so) is the one which non-sociologists sometimes find very difficult to accept. However it remains for us a guiding principle. But it is not only the basic assumption of sociology which sometimes seems to be a closed book to non-sociologists: its methods can appear to be less than rigorous. The experimental method is generally unavailable and consequently the sociologist is confined to direct or indirect observations. Moreover the benefits derived from sociological investigation are far from obvious and often controversial (Elias, 1978; Neustadt, 1965).

Sociology is a broadly-based discipline and at first sight the major questions of the nature of social change and development may not seem especially relevant to engineers. We will argue later in the book that the broad sociological concepts concerned with change and continuity are relevant to engineers but for the time being we merely point to one specific branch of sociology as a kind of *hors d'oeuvre* which has immediate obvious relevance to engineering, namely industrial sociology and the related fields of the sociology of occupations and of organizations. Here sociologists focus on problems central to the work of many engineers. For example the effects of economic cycles on firms and organizations and the organization of work in terms of its structure, and effects on individuals are areas which are well developed in the literature (Watson, 1980). Other topics extensively discussed by industrial sociologists are the legal and social constraints which impinge on work organizations (Eldridge, 1971). Another central concern is the nature of power and conflict at work, along with the related issue of control of work-place behaviour (Hill, 1981). The nature of the bureaucratic organization of larger employing units and the effects on the people who work in them have also generated considerable interest (Child, 1984). Management attitudes and structures of production systems and influences on productive performance, and the causes and consequences of technical developments have all been extensively investigated (Littler and Salaman, 1984). Payment systems, absenteeism, skills, leadership and supervision, decision-making and communications at work are also all grist to the industrial sociologist's mill, as are worker and management beliefs, ideas, values, job satisfaction and behaviour (Rose, 1975). The final aspects of industrial sociology we will mention here are the various factors external to the work situation which have been found to impinge directly on employee behaviour. Sex and gender, education, home life, local communities and age (Brown and Harrison, 1978; Brown, 1976; Goldthorpe *et al.*, 1969), have all been identified as crucial in understanding worker behaviour.

We suggest that this is a source of information, for the most part unknown to or untapped by engineers.

Thus, many of the topics which concern individuals who organize productive work are precisely those which have interested industrial sociologists. Yet the interchange of ideas between sociology and engineering remains limited. This may be due to the fact that sociology appears to lack coherence or that there are few obviously cumulative research findings. The answers may be even simpler: the sociologists' findings may not be widely enough available to engineers or many people, not just engineers, may be suspicious of the sometimes rather private language used in sociologists' texts.

Notwithstanding these difficulties we take it as axiomatic that industrial sociology is, and should be, relevant to practitioners and students of engineering. The subject matter of sociology *is* broadly about the world engineers inhabit. Engineers deal with human as well as physical resources. Human resources in this sense are the very stuff of industrial sociology. An appreciation of the social and political as well as the economic settings in which engineers work is potentially highly useful. This does not of course imply that sociology itself has all or even very many of the answers, but it is surely reasonable to argue not only that it might, but also that it ought, to provide some.

These ideas are not especially original in the context of engineering education. Western European and North American engineering programmes traditionally contain a significant element of what we have defined as industrial sociology. Indeed management-type problems are treated as part of the engineering dimension or process. Furthermore, in Britain there have been influential advocates of very broadly-based engineering courses (cf. Wild, 1982). However the Finniston Inquiry found that where broadly-based elements were included they tended to be tacked on to the supposedly 'real' engineering component and were sometimes tarred with a liberal studies brush. Here we are aiming to undermine the traditionally limited British view of engineering education's potential.

Our Audience

This book is intended for use by engineering and comparable undergraduates, but we hope it will also attract a wider readership amongst practising engineers, sociologists with an interest in engineers and engineering, and all those interested in Britain and in engineering – industrialists, politicians, educators and civil servants – indeed anyone concerned with the future of engineering and the British economy.

There are several reasons why we consider the production of a sociology book for engineering students to be desirable. Most obviously the teaching of sociology to engineering undergraduates in Britain is a fairly new but growing phenomenon. Hitherto much engineering education not consisting of mathematics and natural science-based subjects has generally been of a liberal studies nature, or has consisted of the uncritical teaching of management studies and economics. The former has apparently been intended to humanize engineering students, who very often realize the fact and reject the material taught as a result. For example they wonder why those who study and teach the self-styled humanities are not taught some engineering and science to counter *their* narrowness; and they feel that the good that engineers do and the often considerable extent of their contact with a wide range of people are not appreciated enough. Management studies and economics are very often packaged and taught in ways which fail to tie in closely enough with the technical details of and the human issues raised by particular engineering tasks.

The sociology which has sometimes been taught has itself manifested similar weaknesses. First, it has tended, at best, to stimulate general interest rather than genuine curiosity, because it has rarely been designed for, or been particularly sympathetic towards, the needs of the engineer. Secondly, it has implicitly reinforced the view that engineering consists of the 'mere' application of natural scientific knowledge. Several existing texts in the sociology of industry are so affected. The response of engineering undergraduates and engineers tends to be variously hostile or indifferent. Hostility is generated because the political or humane concerns of sociological writers patronize the engineer, who is assumed only to be a self-interested technical organization man. Indifference is however more usual since the sociology is thought to be largely irrelevant to the activity of engineering.

The book is aimed at all types of engineering undergraduate and engineer, although we expect it will be of more interest to those oriented towards or in manufacturing rather than to those who are to enter or who work in construction, although we do not entirely neglect the latter. We have tried to avoid the use of sociological jargon, which is increasingly rejected by sociologists anyway, and to write a book which says more about society than sociology, so that the book can be used by engineering teachers as well as by sociologists who teach engineers. Additionally the book ought to interest sociologists who want to learn about engineers and engineering, irrespective of whether they themselves teach engineering students. It should also be of interest to undergraduate and postgraduate students of other subjects apart from engineering, such as those reading business studies, management, and the social and natural sciences.

The Plan of the Book

The book is divided into two parts. The first defines the problem with which we are concerned. This is the marginal status of engineers and engineering in British economic, political and academic life. Our starting point in Chapter 2 is to highlight the crucial role of engineering in the development of human societies. We look at the ways in which sociology originally analysed this. The contributions of classical sociology (*c.* 1800 to *c.* 1920) are therefore examined.

The most important contribution to understanding the centrality of the productive process to social development is found in two ideas central in classical sociology: *homo faber* and the division of labour. *Homo faber* represents a socio-psychological model of man which stresses that the most important human characteristics are productivity and creativity. It is these characteristics which give our species its identity. A commitment to this view of human nature places engineering in a pivotal position in human society, it being the activity primarily concerned with making and doing things productively and creatively. The concept of the division of labour refers to the fragmented and divided nature of productive activity and social life itself.

In the next chapter we consider the economic context in which engineering operates in Britain and focus in particular on the so-called British economic problem. The links between the British economic problem and Britain's treatment of engineers are drawn out. The relationship between the historical distaste for engineering in Britain and British attitudes to manufacturing and industry is outlined. We also consider some of the apparent explanations and proposed solutions offered by politicians, commentators and various other pundits to the problems. We find the solutions in the main to be concerned with symptoms, not causes, and therefore not very sound solutions. In order to develop an argument, in Chapter 4 we move away from a consideration of the internal problems of the British economy and examine its broader international context. In Chapter 5 we consider the legal and political context of engineering, drawing some comparisons with Britain and elsewhere. The role of engineers in political life and culture is our main focus of interest. This is analysed with respect to the power structure of British society.

With our problem defined as the disjunction between engineering's productive importance and its lack of economic and political power, in Part Two of the book we develop sociological explanations of this. Our central concept is the division of labour in industrial-manufacturing and related processes and systems. In Chapter 6 we consider engineering education and socialization into engineering careers. Our analysis in this chapter is both structural and experiential. Thus we are concerned not only with the formal mechanisms whereby people become engineers but also with what it means for the person going through the process. We suggest that the meanings of the experience derive from broad socio-cultural factors. In particular the impact on engineering education in Britain of the values of *laissez-faire*, and a bias towards intellectual rather than hand skills, are targeted as major explanatory variables. The ways in which these values continue to pervade engineering education, even in the post-Finniston era, is considered. Chapter 6 concludes with a consideration of the social-psychological consequences of these socio-cultural variables on persons entering engineering. Chapter 7 deals with theories of motivation at work. We examine the major perspectives and theories on work motivation and job satisfaction. We attempt to deal with the ways that socialization into engineering affects motivation. Chapter 8 begins the discussion (continued in Chapter 9) of the work context of engineers and engineering by looking at the network of relationships in which the professional engineer works and considering some of the typical non-engineering specialists and others with whom engineers have to deal. Chapter 9 discusses the organization and experience of engineering. This chapter is also concerned with structural and experiential aspects of the issue and deals both with data concerning engineering work (where it takes place, which sectors of the economy and which type of economic activity) and with engineers' experiences of these things. Chapter 10 examines the collective response of engineers to their position in the division of labour. It highlights the inappropriateness of many of the collective responses. These responses, we argue, have tended to amplify and mirror the existing problems facing engineers rather than solve them. In Chapter 11 we apply the sociological and economic concept of the division of labour directly to engineering work in order to explain in sociological terms the problems raised in the earlier chapters. This is the critical chapter, in the sense that the concept of the division of labour is used to draw together and make sense of a range of disparate information. In the final chapter we spell out our major conclusions and discuss some of the implications of our argument.

2 Engineering and the Social Process

Introduction

In this chapter we have two aims. First we make the point that engineering, broadly defined, is central to human life. Second, we emphasize the fact that important insights into understanding the centrality of engineering can be gained by using two concepts, *homo faber* and the division of labour. These two concepts are not, of course, in any way exclusively sociological, but they occupy a key place in the armoury of sociological ideas. They are especially appropriate when considering engineering. *Homo faber* and the division of labour were central ideas in the early development of sociology. In this chapter therefore we explore the concepts of *homo faber* and the division of labour in human societies and their origins in early sociological writings.

An Engineered World

Britain, the USA, Japan and most Western European countries are generally understood as being very prominent among the advanced industrial societies. Many pundits, including economists, engineers, historians, philosophers and sociologists have produced definitions of industrial society. According to Giddens (1981: 141) an industrial society is characterized by industrialism, namely the 'transfer of inanimate energy sources to production through the agency of factory organization'. In this kind of society machines are systematically made and used on a large scale and these machines use non-human, non-animal sources of energy.

In terms of the history of mankind, industrialization is very recent, starting to be effective only in the eighteenth century AD. The earliest human societies relied primarily on the gathering of food and the hunting of animals. This indeed was the predominant type of social organization until about 12,000 years ago. Tools and weapons were not made of metal until around 4000 BC, the plough was not in use until about 1,000 years later, and iron tools and weapons were not used until roughly 1000 BC (Lenski and Lenski, 1978). This gradual change is sometimes broken into periods which are named after the materials used or the artefacts produced in them: the Stone, Bronze and Iron Ages, and more recently the Steam, Jet and Atomic Ages are typical examples. Man's history has been inextricably bound up with that of engineering, defined broadly as the making of contrivances as aids and adjuncts to life.

Most observers agree that in the early period of man's existence the pace of change was very slow, but with each new device, and each small development in the tools which generally improve human comfort, the rate of change quickened (Bell, 1968). Man's brain, and his body, have changed little during recorded history, while his cultural evolution has been relatively rapid. It was because of the development of more difficult and specialized techniques, in farming and tool-making and tool-using, that towns began to be erected and lived in.

As human societies became richer, more complicated and less dominated by natural forces, social habits became a more powerful feature of their lives and language and myth, both written and spoken, became more central to development. Gradually a few people who could read and write began to be distinguished and usually elevated above the makers and most of the doers. Farmers and religious men, as those in charge of the written records, were normally closer to rulers, landowners and generals than the artisan types, and consequently had more say in the ordering of affairs (Bloch, 1962). Yet the fact remains that engineering by various names has been at the root of much social, economic and political development.

The technical accomplishments of human societies and the reliance of human society on tools and the use of tools is fundamental. From the earliest societies where the only technical adjunct to life was the simple wooden spear or stone axe, through the development of the bow and arrow and horticultural implements, the use of metallurgy and the development of the plough, to highly complex industrial society, the dependence on technical help for human existence has been a central fact of life. The question arises as to who are and who were the men and women who initiated, developed and tinkered, to bring about technical developments? It is that group which today we recognize as a separate occupation – engineers – but in an earlier era might have called mechanics, artisans or practical people. The search for solutions to problems of existence in a practical rather than philo-sophical way has always been the role of the engineer in this broad sense. The creativity, activity and practical initiative are those parts of human nature expressed as *homo faber*.

Along with technical development went social development, and the increasing complexity of social systems is what the division of labour describes. The notion of the division of labour articulates the fact that human life is and human development has been characterized by increasing specialization of tasks, roles and functions, a specialization made possible by the wealth produced by the technical developments of *homo faber*.

We argue that throughout the long transition from a predominantly agricultural to a predominantly industrial society, engineers (broadly defined) played a crucial role. Yet surprisingly, in much of the historical and sociological literature this has received relatively little attention. The place of engineers in history has been neglected by many humanists, including historians who have tended to emphasize instead the role of 'people like themselves (scientists and other intellectuals), and they have . . . largely disregarded the messy technology that has been associated with virtually every important historical change and which critically impinges upon every man in his day to day life' (Smith, 1970: 493). As Smith hints, this neglect is partly related to the fact that natural scientists have, in some countries but not all, claimed at least some of the glory of modern engineering's achievements for themselves (Fores, 1977; Klemm, 1959). In the final analysis far more has been written about the history of engineering than about engineers, 'and much more about those who promoted engineering works, than about those who actually designed and built them' (Rae, 1975). This neglect is not however true of the writers who are now regarded as the classical sociologists who, although they may not all have called themselves sociologists, were writing in either a recognizably sociological way or produced ideas which have become part of the sociological tradition.

Engineering and Work in Classical Sociology

There are a number of key figures in the development of sociology. Here we consider some who addressed engineering-related issues: Saint-Simon, Marx, Durkheim, Weber, Veblen, Comte, Spencer and Pareto.

Saint-Simon (1760–1825) was a liberal French aristocrat who fought with the French Army against the British in the American War of Independence, who became a radical republican and an architect of socialist thinking in the French Revolution, who was an influential champion of industrialism, of engineers and bankers, and who saw the future as an age of the machine and of peace and progress. His ideas were later called 'the religion of the engineers' (Hayek, 1955). Saint-Simon's primary concern was with the nature and effects of industrialization, which were also central to the thought of Marx, and Durkheim and Weber. He felt that it involved progress to a scientifically-based kind of social order. Saint-Simon suggested that the future would belong to a democratic, one-class society consisting of *les industriels*, by whom he usually meant all those involved in manufacturing production, as opposed to parasitic types whose loyalties related to the feudal past. Men would conquer nature rather than each other. Rewards would go to ability and effort, and inherited social privilege would disappear. Government would be concerned mainly with the productive use of resources, rather than with domination of the weak.

One of the giants of nineteenth-century economic, political, philosophical and social thought, Karl Marx (1818–83) was keenly interested in the way work organization affected other social relationships. The idea of *homo faber*, man the maker, creator and doer, best describes Karl Marx's attitude to productive work. Marx argued explicitly that it was the essential human condition that men could first imaginatively construct their world, and then physically produce their environment through work and through such work could fulfil themselves. These two characteristics of imagination and work were, Marx believed, the uniquely human qualities which distinguished men from animals. The ultimate tragedy of the history of mankind, according to Marx, was that the realities of the organization of people for work, particularly its exploitative nature, meant that few people did actually achieve fulfilment through it. This condition was termed alienation.

Alienation arose because the propertyless worker under capitalism owned neither the means of production nor the results of his work; he was 'degraded to the most miserable source of commodity'. The term 'alienated' literally means cut off, separated or estranged. Alienation for Marx had four main aspects. First the worker was alienated from the *product* of his labour. Second, the worker was alienated from the act or *process* of production. Work is forced labour designed to meet the needs of others. Third, the worker is alienated from his *species-being*, from his own human nature, because his work is only a means to the end of subsistence rather than a source of creative self-development. Finally, and as a consequence of the former, men are alienated from *each other*: the system of divide and rule makes relationships between men calculative, selfish and untrusting. Under capitalism, higher wages could only produce better-paid slaves, and a happy or 'satisfied' worker who accepted 'the system' would only be a kind of 'cheerful robot' whom it had 'bought'.

After alienation the second main strand in Marx's work concerns social development and the division of labour, whereby forms of ownership and of production changed through history. Early tribal societies had been replaced by 'ancient' ones (as in Greece and Rome), which were replaced by feudal ones. Feudalism was replaced by capitalism, which would then be overthrown by socialism. Socialism would then evolve into communism, and this change is quite well expressed thus: 'from everyone according to his abilities, to everyone according to his needs'. Each of these different stages of social development represented different ways of organizing productive activity. The whole of Marx's work is therefore premissed on an analysis of the nature of production, and the technical means whereby such production is engineered (Avineri, 1968; Zeitlin, 1967).

In the work of Emile Durkheim (1858–1917), the importance of the organization of work is also evident. Durkheim was a major intellectual figure in France for about twenty years: he was 'Leftist' – meaning anti-clerical, republican and progressive rather than socialist. He was consulted by governments about moral education, and he succeeded in getting sociology established as a school subject. His concern with social order and harmony reflected the relative lack of it in France.

In *The Division of Labour in Society* (Durkheim, 1933) he distinguished between two types of solidarity – the feeling of people that they belong together and the fact of their doing so. Mechanical solidarity, which brought people together in a total way, was typical of primitive, ancient and medieval societies and of many present-day family relationships. Organic solidarity, typical of modern societies, was far more complex. Here solidarity was not based on a simple feeling of belonging together, but on a complicated system of contractual relationships. Mechanical solidarity prevailed when the division of labour was simple and social ties were secure and perhaps even all-encompassing. Organic solidarity was based on a more complex division of labour, when ties between people are partial and commit them less. With organic solidarity relationships were often very specific and limited and easily broken, like those that exist between passengers on commuter trains. Contractual relationships under organic solidarity did involve mutual rights and obligations, but they referred to the subject matter of the contract and nothing else: they were associated with loneliness. Societies held together by mechanical solidarity emphasized fellowship and faith, whereas reason and the rule of law were more prominent under organic solidarity.

For Durkheim the central social problem of modern times was not Marx's alienation of the proletariat of propertyless workers. Instead it was anomie, the state of an individual or group which feels deprived of relationships, that it does not belong. Under mechanical solidarity even the village idiot had a place in the scheme of things, but the fragmented, perfunctory, transitory, often unreliable and apparently meaningless character of many relationships under organic solidarity made individual existences seem much more isolated. Individuals lacked a permanent, comprehensive context for their lives.

Mechanical solidarity had its problems too, however, because it integrated people into communities at the expense of individuality, often using repressive laws. It was typical of societies further back along the scale of social evolution than those typified by organic solidarity. Once it began to break down, as societies began to become more mobile and complex, anomie resulted, as the price to pay for change. A kind of reintegration was necessary in order to be free of anomie. This meant that social integration, linking individuals while allowing them to grow as individuals, had to coincide with system integration, whereby society can be mobilized completely yet with each of its sub-systems having some freedom of movement.

For Durkheim no particular society completely exemplified one of the forms of solidarity, because all societies manifested both along with a certain amount of anomie. Typical results of anomie included moral anarchy, unbridled conflict, unlimited desires or feelings of meaninglessness, some of which clearly exist in a late twentieth-century Britain which having lost its empire failed to find a satisfactory role to replace its imperial one. If social regulation of the division of labour was excessive or 'forced' as opposed to 'anomic' (or Durkheim's ideal, 'spontaneous'), the results would include an atmosphere of repression, and of fatalism or smouldering resentment. Ideally a moral consensus should moderate competition and conflict in society spontaneously, so that a middle course could be steered between anarchy and repression (Durkheim, 1933, 1952).

A near contemporary of Durkheim was the German Max Weber (1864–1920). His writing defies any kind of simple summary, ranging as it did over philosophy, history, religion and culture. Moreover a lot of what Weber wrote was an attempt to refute or refine some of what he perceived to be the excesses of Marxian thinking. However throughout much of Weber's work a concern with work and its organization, and in particular the gradual rationalization of society as he saw it, forms an important unifying theme.

Weber argued that domination, whereby the ruled saw it as their duty to obey their rulers' orders, took three main forms (Weber, 1947,

1948). *Charismatic* domination meant that more or less blind faith in unique qualities of the leader – a demagogue, hero or prophet – legitimized his power. *Traditional* domination, historically the most common type, meant legitimation by and obedience based on custom. Under *legal–rational* domination, which Weber felt was typical of 'modern industrial capitalism', power was legitimized by a belief in the rightness of the law, and exercised within legally-sanctioned limits. Means were rationally related to ends, and proper procedure was central. Whereas charismatic domination usually meant loose, unstable administration, legal–rational domination meant impersonal bureaucracy. Weber warned against believing in some kind of inevitable historical trend from charismatic or traditional to legal–rational domination. For Weber rationalization meant amongst other things the bureaucratization of administration. Thus legal–rational authority systems were most typically manifested in impersonally-bureaucratic administration, with leadership based on and constrained by legal–rational principles.

Weber's writings on bureaucracy have probably been the major starting point for the study of work organization, which has grown massively since the 1940s. Debates with Weber's ghost about bureaucracy and work organization have informed many study areas including employee motivation, decision-making, industrial conflict, management styles, job design and leadership. His ideas remain important because variants of bureaucratic administration seem to be the norm in large employing units in most of the developed countries, both capitalist and socialist, and to be a significant feature of modernization elsewhere.

One American writer who gave prominence to engineers was Thorstein Veblen (1857-1929). Regarded as something of a cult figure and oddball, Veblen wanted engineers to play a more important and powerful role in industry and social life than the European classical sociologists did. He felt that the power of finance and of big organizations had subverted the position and the creative role of American engineers, so that they had become the servants of the powerful financiers who they had helped to bring into being. Veblen was more optimistic than most sociologists have been about the role of engineers. He predicted the emergence of a technocracy, with the engineer in charge as the guardian of the community's material and social welfare who, with dispassionate and rational attitudes, would impose purpose and coherence on a hitherto chaotic and irresponsible economic and political scene.

Temperamentally Veblen was a debunker and a rebel. His interests were extremely wide-ranging. He wrote about idustrialization, capitalism and alienation, women's place in society, international and academic politics, and the lifestyles and leisure patterns of the wealthy. Diggins (1978) attributed his neglect by most sociologists to the diversity of his interests, his apparent lack of seriousness, his somewhat unusual (even eccentric) values and behaviour, and his originality which set him apart from other academics.

Veblen, his background and his ideas have always seemed a little marginal both to American society and to European social thinking, although both influenced him considerably. He was critical of capitalism, yet influenced by ideas about social evolution which saw its growth as part of the natural order of things. He was both an admirer and a critic of Marx. He criticized and complemented Marx's emphasis on productive labour as capitalism's major source of wealth by arguing that technical development was also playing a central part and that the importance of intangible forms of property was growing. He was very American in his regard for the 'plain man' who, partly in spite and partly because of his simplicity, outperforms and outwits his social betters.

Veblen's thinking about alienation was rather different from Marx's, partly because he saw much more continuity and, in a way, complexity in history than Marx had. He attributed it to the basic and immemorial emotions of pride and imitation, rather than to a mere desire for economic gain. Thus men had long seized and exploited property (especially women) in order to assert their dominance. Alienation happened whenever individuals subordinated their 'instinct of workmanship' to a desire to display prowess and to become socially dominant. Its origins lay in the elevation of aggression and greed to honourable status and to the degradation of useful labour to irksome and humiliating activity.

Exploitative economic relationships came to dominate other areas of life through a 'psychology of emulation' of 'commodity fetishism' and 'conspicuous consumption'. The insatiable pursuit of goods allowed consumers to display prowess and to identify with the dominant predatory class, and in doing so it helped to integrate society. Such ideas are clearly powerful, plausible and attractive, and they were quite original too.

Veblen applied his ideas to the wealthy, upper, 'leisure class' of the USA in the 1890s in the so-called Gilded Age of American capitalism. These people were too wealthy to need to work, and their conspicuous consumption of accumulated capital was the main symbol of their predatory success. To maintain their honour they had to be seen not to work, to consume time conspicuously and unproductively by a display of 'knowledge of dead languages and the occult sciences; of correct spelling; of syntax and prosody; of the various forms of domestic music and other household art; of the latest proprieties of dress, furniture and equipage; of games, sports and fancybred animals . . . manners and breeding, polite usage, decorum, and formal and ceremonial observances generally' (Veblen, 1953).

He applied the same analysis to engineers (Veblen, 1921). They were part of productive industry, which non-productive business, in the form of finance capital, lived off. Significantly the conspicuous consumption of technical knowledge was usually only prominent in wartime; in peacetime the less useful arts were more visible.

Veblen's thoughts on technical and social development, especially with reference to France and Germany, are also quite significant. Technically advanced societies with old-fashioned institutions he termed 'unstable compounds'. Veblen also wrote tellingly on academic pretence and bureaucracy in the USA. But his most powerful thoughts were those on emulation, whereby alienation was linked to consumption rather than production. It is doubtful whether anyone has explained the 'paradox of labour', whereby those who create wealth can be despised, better than he.

The Frenchman, Auguste Comte (1798–1857), was Saint-Simon's secretary for a short time towards the end of Saint-Simon's life, and his disciple, but eventually broke with him. Comte invented the word sociology, and although his work is now largely ignored his contribution is of interest to us. It is in the analysis of what Comte called social dynamics that the role of the technical expert was seen as crucial. Comte believed in a progression, both of our understanding of the world and of human societies, from forms of existence dominated by religious ideas through the forms dominated by the discovery and application of scientific laws. He believed that sociology was the highest form of scientific reasoning, which would ultimately lead to the ability to control social existence itself. The predominant form of political organization in the religion-based type of society was military, in Comte's view. This was to be replaced, in scientific or positivist society, by industrial organization. Thus the people at the heart of the industrial complex would be important politically, although subservient to sociologists, in the new social order. Above all, he associated industrial progress with peace and order.

Herbert Spencer (1820–1903), a self-taught English philosopher and sociologist, applied the biological idea of evolution to social development. He coined the term 'the survival of the fittest' and his so-called Social Darwinism was applied to events within and between societies. Like Saint-Simon and Comte before him, Spencer had great faith in industrial progress and industrial societies. He associated them with individual freedom, justice, peace and wealth. He was particularly concerned with the way in which, in growing industrial economies, the division of labour became more elaborate and differentiated internally: like plants, societies ascended the evolutionary ladder. Here he was an important influence on Durkheim's (later) thinking on the division of labour. Like Saint-Simon, Comte, Marx and others, Spencer regarded the 'captains of industry' as the power-holders of the future.

The Italian Vilfredo Pareto (1848–1923), unlike the other writers we have considered, was himself an engineer (and railway company director). Like Comte he treated the study of the human mind as an integral part of social analysis. Like Marx, he saw economic and social arrangements as inextricably linked. Indeed he is probably better known as an economist than as a sociologist. He did not so much extol the virtues of engineers and engineering in his sociological and other writings, but rather used ideas and metaphors from engineering in his analysis. In particular, the ideas of system and of mechanics were central to his thinking. He is principally remembered in contemporary sociological theorizing for his contribution to the study of elites, arguing that power circulates between the 'foxes', who rule by guile, and the 'lions' who rule by brute force. His idea of system was very influential in the Human Relations movement (see Chapter 7).

Although it is only really in the work of Marx of those we have looked at, that the concepts of the division of labour and *homo faber* are brought together explicitly, the notions of social division, social differentiation and the importance of human action through creative activity is addressed by all of these writers. They established the ground rules of what was to become sociology. The notions of structure and action, of man living in a structured and socially differentiated environment and acting in and on that environment, were laid by these early writers. They, in varying degrees, acknowledged the importance of engineers in these processes. As we will argue elsewhere in this book, while the ground rules of structure and action have remained intact, the centrality of the engineer in all this has too often been lost. Amongst other things we wish to help re-establish that centrality in this text.

Conclusion: Homo Faber and the Division of Labour

Man lives in a world which is shaped and created by his own activities. The world which man creates is one in which human roles, tasks, functions and relationships are divided and separated from each other.

These divisions comprise what sociologists call the structure of society. The division of labour is the product of human creativity and purpose as embodied in the idea of man as *homo faber*. We suggest that engineering plays a crucial role in this.

Homo faber is an important idea because it helps to explain how man's culture, his tools, habits, traditions, beliefs, skills and abilities have developed. *Homo faber* helps to highlight the fact that man is the only really self-made animal and the significance of man's culturally transmitted characteristics which distinguish him from other animals, whose attributes are inherited to a much greater degree (Lorenz, 1937; Tinbergen, 1951, 1968).

Although people have always worked, and always been *homo faber*, the nature of work has changed greatly over the centuries. Economic development has made much work physically and often mentally less demanding, and often a lot less time-consuming. The division of labour has continually fragmented, reorganized and transformed work and working relationships. Higher living standards have helped to make government more democratic and to raise expectations about work and employment. Yet in spite of very high living standards for the wealthy in a few parts of the world, the vast majority of people still have to work in order to live, with a large minority still extremely poor. For this and other reasons work looks likely to remain central to life in the foreseeable future.

Berger (1964) argued that work could be experienced by people in three main ways. It could be life-enhancing, even all-absorbing, and the major source of a person's identity (as for many professionals, artists and craftsmen). At the other extreme it could be variously dirty, dangerous or menial and a source of suffering and a threat to one's identity and self-esteem (for example, coal mining, office cleaning, and refuse collection). Between these two poles there lay a grey area of jobs which produced neither fulfilment nor suffering in people: they were tolerable without offering much satisfaction (for example, bank officials, semi-skilled factory operatives and junior civil servants). Berger and others have argued that the last category has been expanding at the expense of the other two.

Berger's categories have been paralleled and developed by Parker (1972, 1976) who was interested in relationships between work and other areas of life. For Parker leisure and other non-work activities typically associated with life-enhancing jobs tended to be an *extension* of work. Examples might include the electronics engineer whose hobbies were hi-fi and home computing, or the social worker who did voluntary work or belonged to pressure groups on behalf of clients. Particularly arduous, degrading or otherwise unpleasant jobs tended to be in *opposition* to non-work activities. Frustrated and unfulfilled workers drew a firm line between work and the rest of life. They either concentrated strongly on the private worlds of family and hobbies, or indulged in such forms of explosive compensation as heavy drinking and using the services of prostitutes. Examples of the former include assembly-line workers in car manufacturing (Goldthorpe *et al.*, 1968, 1969); the latter include deep-sea fishermen (Tunstall, 1962). Finally the 'grey-area' jobs which were neither all-absorbing nor degrading were associated with a pattern of *neutrality*. Work and non-work were simply different parts of life, complementary and each with its own rewards and problems.

Is there any general theory or set of ideas about human nature which can explain, among other things, why and how people work? The short answer is yes, and we spell elements of it out below, but it suggests, almost paradoxically, that to understand the details of particular circumstances and cases is often much more useful than to think in general terms. We know that men, unlike animals, can develop languages and employ abstract thought, and that men have very superior abilities to evaluate alternative ways of behaving and to control and channel their bodily urges and emotions. As *homo faber*, men try progressively to conquer and organize the world, simultaneously tool-making and tool-using, searching for individual fulfilment and social welfare, aggressive and cooperative, and acting logically and rationalizing before, during, and after events.

The social institutions which men create vary in terms of power, influence and success. Hierarchy and domination are present in some form in virtually all societies. They tend to be unstable because all power tends to provoke resistance and because organized, goal-directed behaviour usually produces at least some unintended and often disruptive consequences. Thus *homo faber*'s fundamentally awkward experimental and playful nature always reasserts itself in the face of attempts to organize existence. Aggression is a normal, natural and fully human part of this: 'we are the cruellest and most ruthless species that has ever walked the earth' (Storr, 1968).

However, self-denying altruism is equally human. So too is cooperation. *Homo faber* is both selfish and selfless. Human societies which emerge out of *homo faber*'s activities involve conflict and cooperation. The division of labour expresses both domination and compliance, superordination and subordination. Our range of behaviour, the variety of our social arrangements and the extent of our physical and mental creativity appear to be virtually infinite.

The main thesis of this book is that the division of labour as it exists in British society often seems to take self-destructive forms. This is particularly marked in the industrial manufacturing system, wherein not only the divisions between development, design, production, marketing and so on are often unsuitably integrated, but the central role of the tool-maker writ large – the engineer – is often undervalued and under-powered.

3 Engineering and the British Economic Problem

Introduction

The division of labour refers to social differentiation. There are many facets to social differentiation; people are differentiated for example on the basis of age, gender, race, regional and national origin, occupation, colour of hair and physical stature. However there are two dimensions of social differentiation which require particularly close attention in respect of engineers: status and class. The reasons for this are, first, that compared with some occupational groups like doctors and scientists, who might otherwise be thought of as in some way equivalent in terms of qualifications held and the complexity and responsibility of their jobs, British engineers do not fare particularly well in terms of prestige, financial rewards and power; and, second, compared with engineers in other countries British engineers do not do well either. Prestige, financial rewards and power are aspects of class and status.

Status and class are terms which are used in very precise ways by sociologists. Status refers to lifestyle and patterns of consumption, class refers to an economic relationship in the productive, particularly market, aspect of human life. Status is historically the older type of ranking, based on the ways in which different 'estates' or status groups had access to the surplus produced in agriculture or hunting. It refers to a positive or negative estimation of honour or prestige given to individuals or groups (estates). It is expressed in ways of living. It is intimately bound up with patterns of consumption and concomitant social and domestic habits. The idea of status tries to capture a range of more or less tangible phenomena from mode of speech, dress and manners, through to patterns of behaviour, morals and beliefs. It may be subjectively felt, through the living out of some kind of life pattern and associated self-image, or it may be bestowed by others. Self-images of status and status bestowed by others need not necessarily coincide, but they frequently do. A person's age, occupation, achievements, racial origins, and place and type of residence can all be relevant in such evaluations. It is convenient to distinguish between *ascribed* status and *achieved* status. Ascribed status refers to inherent qualities such as birth, age or skin colour. Achieved status is a consequence of personal effort or the appearance of it, resulting in the occupation of particular social roles, for example worker, farmer, businessman, husband, engineer, doctor and so on.

Status is generally conferred in ways which legitimize existing differences in power and wealth. Many clerical and other non-manual workers for example feel that they ought to be paid more than skilled manual ones, simply because they themselves do not 'work with their hands'. Depending partly on our backgrounds we might assume that poorer people are lazy spendthrifts, or saints, or simply people with no incentive to work. More fundamentally, the values of the powerful and successful tend to colour all our tendencies to honour or to despise others. Thus when landowners are politically powerful, industrialists tend to lack status relative to their wealth, and vice versa.

Class is *not* the same thing as an occupation, and is *not* defined by income. Class *is* an economic phenomenon determined by market relationships and may therefore be related to occupation and pay, but class and occupation and pay are not coterminous. A class is defined as a group in a competitive market relationship for land, labour or capital; the factors of production. A competitive market relationship in this context is one in which the factors of production, land, labour and capital are treated as commodities having value and price. Competition arises because the factors of production are more or less scarce. Classes are formed in the process of economic exchange of the commodities. Different groups of people (classes) enjoy differential access to and control over the various factors of production and therefore differential power in the process of exchange and competition. Consequently, different groups (classes) enjoy differential access to the wealth produced by the bringing together in these market relationships of the factors of production. Classes, and a society differentiated on the basis of class, cannot exist without market relationships. Those with or without possession or control of particular factors of production form the particular class groupings. (For further discussion of class and status see the introductory texts mentioned above and also Giddens, 1981.) We consider the class position of engineers in detail in Chapter 11. Here we consider the question of status.

Measured in status terms, British engineering does not, as an occupation, rank very highly, even though some individual engineers do. There is a considerable literature on this topic. Watson (1976) reviewed numerous studies of occupational prestige in most of the industrial countries, and concluded that engineers were indeed less highly regarded in Britain than in any of the others. In the mid-1960s, British professional engineers ranked below doctors, solicitors, university lecturers, research physicists, company directors, dentists and chartered accountants, and only just above primary-school teachers. In the USA engineers were rated roughly equal to scientists and above architects, lawyers and accountants. Australian, Canadian, French, German, Russian and Scandinavian data all suggested that engineers are ranked higher in those countries than in Britain or the USA. Engineers in the 'developing countries are firmly and consistently placed towards the apex of their professional complex'. Britain produced far more science graduates relative to engineers than was the case in other industrial countries where engineering courses also attracted more of the most able students. Natural and social science as well as medicine were all much more highly regarded than engineering in Britain. Since Watson wrote the prestige of social science has declined but other relevant changes seem to have had something of a mutually self-cancelling quality. However, the sum of all changes in educational choices and content, in choices of careers, in occupation formation and occupational organization, is probably slightly beneficial for British engineering's long-term future, on balance. Even so the problem remains both widespread and deep-rooted.

For example a survey conducted in 1978 for the Finniston Committee of Inquiry into the Engineering Profession asked a sample of Britain's general public a number of questions about engineers and engineering. Engineers were rated below doctors and accountants in terms of career prospects and in doing this the members of the sample seemed to equate engineers with draughtsmen/women. People from the lower social classes rated engineering careers more highly than people from the higher social classes. There was some evidence of prejudice against engineering as a career for women especially on the part of female respondents and those of lower social class. Respondents were asked to identify the kinds of work 'an "Engineer" ' might do, and 68 per cent referred to manual-level jobs with a mere 13 per cent associating the word 'engineer' with a professional type of job. Nearly a fifth of the respondents seemed to have very little idea of what engineers do. However about half of the respondents did recognize that professional engineers either did work of a professional kind or were people with higher-level qualifications than 'ordinary' engineers or did work which involved some kind of managerial responsibility.

After being given examples of professional engineering tasks, respondents were then asked about the kinds of qualifications needed by professional engineers. Few realized that professional engineers generally need degrees. A quarter and a third thought, respectively, that electrical and mechanical engineers needed to have served an apprenticeship. At least 38 per cent thought that a lower than HNC/ HND level qualification was all that was needed. Clearly a lot of ignorance was present although it is likely that most people would have shown similar levels if not types of ignorance about qualifications needed by such people as doctors or accountants. On pay, people from the higher social classes felt that engineers were not very well paid given their qualifications and responsibility levels. People whose households included an engineer had less rosy than average perceptions on engineers' pay. A small majority (53 per cent) felt that it would be a good idea for their children to aim for a career as a professional engineer *after* they had been told the sort of jobs that professional engineers do: 37 per cent were negative on this score. Overall these findings clearly confirmed the view that British engineers lack status. It would for example be astounding if 68 per cent of the population identified accountancy or dentistry with manual work (even if the latter is, and an unpleasant kind too!), or if 37 per cent felt that careers as accountants or dentists would not be ones that they would encourage their children to follow.

Also in the late 1970s Swords-Isherwood (1979; cf. also Glover 1978a) briefly but effectively reviewed a wide range of evidence on the backgrounds and careers of managers in Britain, the USA, West Germany, France and the USSR. British managers she characterized as coming from mixed/low social class backgrounds whereas her conclusions for the other countries were: USA – mixed/high; West Germany – high/mixed; France – high, and the USSR – high. On the percentages of graduates in management and in the general population her conclusions were: Britain – low/low; USA – high/ high; West Germany – high/low; France – high/low; and the USSR – high/medium. On managers' main undergraduate subjects of study the conclusions were, with the more important subjects first in each case (except France and the USSR where only one was referred to): Britain – liberal arts, science, engineering; USA – liberal arts, engineering, science; West Germany – engineering, business economics, law; France and the USSR – both engineering. For all the countries, engineering was very significant in postgraduate courses studied by managers but in Britain and the USA the so-called management subjects like accounting were more prominent. In Britain, unlike all the other countries, part-time educational routes had been important as sources of qualifications and jobs. The likelihood of being promoted from the shop floor was high in Britain, medium in the USA, low everywhere else. Swords-Isherwood's conclusions fit well with the more general one that engineering in Britain and the USA lacks prestige compared with engineering in Europe, but that the situation is not so 'bad' in the USA as it is in Britain.

Another type of evidence on the status of an occupation is that on choices of occupations and of types of specialization within occupations. In Chapter 6 on engineering education we explore in detail the point made above about British engineering degree courses tending to attract less able 'science-orientated' school-leavers than natural science degree courses. But it is also worth noting that even within engineering, prejudice against manufacturing affects choices of types of work to be done after graduation. Studies of final-year British engineering students' choices of functional types of work conducted in the late 1970s show this quite strongly (Herriot, Ecob and Hutchinson, 1980; Herriot, Ecob and Glover, 1981).

Research and development and design were more often the students' intended choices than 'dirty production or sordid sales'. Many of the students appeared to feel that 'significant others' like lecturers, close relatives and friends would prefer them not to work in the latter areas. Students' choices of employers often depended on whether they felt employers would offer them opportunities to do research and development and they tended to prefer large private-sector organizations as potential employers for this reason. Their beliefs about the outcomes of particular types of work fitted the overall pattern as follows. 'Technical interest' was thought most likely to come from work in research and development, followed by design, production and then sales. The order of preference for 'training and experience' was design, followed by research and development, production and sales. For 'varied work' it was design, research and development, sales and production. For 'dirty and dangerous work', a 'negative' category, production came first, followed by research and development, design and with sales a long way behind the rest. 'Contact with people' was felt to be most likely to be found in sales, then production, then design, with research and development last.

Although a reading of these last sentences may suggest that the broad conclusions from the study described previously are dubious, a detailed reading of the statistical data shows that they are not. Research and development was the most favoured choice overall, design seemed to be attractive because of its links with it, and unfavourable attitudes to production were both common and strong. Interestingly, a study of mainly large engineering employers' perceptions of engineering graduates' qualities which formed an early part of this research mirrored the main findings. The employers (in 1977 and 1978) generally reported problems in recruiting enough graduate engineers of the right quality and type, especially for production and related areas, compared with research and development. They felt that students, especially full-time ones, lacked knowledge of industry. It was sometimes hard to get graduate engineers to move out of research and development or into sales and marketing. Engineering graduates tended to be less ambitious than others, notably accountants. Course content was criticized by most employers for being too academic in emphasis; they did not want students to learn less science but they did want them to be less ignorant of processes and products and of commerce and finance (Glover and Herriot, 1982).

The paper by Swords-Isherwood (1979) described above suggests that British engineers fare rather poorly in employment. This would indeed appear to be the case. They appear to be employed, not as potential senior managers, but as technical specialists whose assumed lack of wider knowledge and social skills makes them unsuitable for promotion to top posts in which knowledge of finance, markets and the general commercial and political environment is needed (cf. Lawrence, 1977). Finance, marketing, and even personnel and research offer easier routes to the top than production and related functions (Glover, 1978a; Melrose-Woodman, 1978; Swords-Isherwood, 1979; Mansfield et al., 1981; Child et al., 1983, 1986; Glover and Garbutt, 1986; Sorge and Warner, 1986; Whalley, 1986). This is not the case in Europe, where engineers clearly do have the relevant attributes, and where engineering and other specialist functions tend to be more closely integrated than in Britain, with far weaker divisions between 'advisers' in specialisms like finance and technical development on the one hand and 'doers' in production and sales on the other, and with production much more broadly-defined and powerful compared with other functions than in Britain.

To elaborate this point a useful list of six forms of managerial controls over employers, developed by Brewster (1986) can be applied to engineers. First, there is 'control over inclusion', which includes recruitment, promotion and termination. Here engineers appear to lose out in Britain by virtue of the idea that most are (only) potential technical specialists, at best a breed of trusted super-technicians, not potential top job holders. Because they are less likely to be promoted into senior posts than, say, accountants or Oxford history graduates turned marketing specialists, they are also generally more likely to be made redundant when their engineering organizations are faring badly. They are also likely to suffer more from direct (and close) supervision and from the short and medium term performance measurements (Brewster's second, third and fourth types of control) which are often built into production and financial information systems, for broadly similar reasons, even if they need to be trusted more than many other types of employee by virtue of their often very elaborate expertise. Long-term performance measurement (Brewster's fifth type of control) in terms of formal (and also in practice partly informal) performance reviews concerned with 'promotion potential' are also clearly likely to be influenced powerfully by employers' existing perceptions of the roles of engineers and of engineering functions. A strong element of self-fulfilling prophecy will be present with senior managements who are generally non-engineers tending to promote people like themselves. Finally, Brewster's last type of control is enculturation. Its purpose is to create an organization in which everyone shares the beliefs and aims of those with power, to turn everyone into 'ICI people' for example. It can be formal (apprenticeships, meetings, courses, etc.) or informal (socializing, forming groups and cliques, jokes, jargon, counselling) but the underlying aim is to get employees to internalize key values and standards. If commercial and financial language and values, rather than technical ones, pervade the culture of top managements as they seem to do in so many British companies then it may be easy enough for engineers to 'belong' but far harder to become one of those who determine the nature of the culture which they and others are expected to belong to.

Given Britain's high population density and lack of many natural resources, the low status of engineering and manufacturing has considerable implications. What colonization, naval power and commerce once secured for Britain now have to be obtained mainly by competence in manufacturing and other engineering sectors. British complacency towards the latter is not new; it worried some people before 1850 (Barnett, 1972). This complacency has many facets, such as a poor average level of technical competence and a lack of trust between management and workers as well as rigid status distinctions at work. Education has often and with a great deal of justification been blamed for a lack of concern with industrial needs, although perhaps a society ultimately gets the education system it deserves, not vice versa. Politicians and civil servants are often accused of related prejudices including technical ignorance although they too are inheritors of the problem rather than its originators.

Four related points help to account for the low status of British engineering: academic traditions peculiar to Anglo-Saxon education, the engineering profession's lack of unity, political indifference, and industry's lack of concern for technical competence in senior management (Glover, 1973). Britain's engineering failures often seem to be magnified, and its successes denigrated. In manufacturing there has been some evidence of conspicuous consumption of engineering and science graduates in misleadingly titled (because they largely did development and design work) research and development departments by employers who did not know how to use them properly (Pym, 1969; see also Jewkes *et al.*, 1969, and Langrish *et al.*, 1972).

On the relatively low status of engineers in manufacturing, some mechanical engineers in one study felt that they had been identified with dirty manual work, or as intellectual 'long-hairs', or as being narrower and less cultured than other managerial-level people (Gerstl and Hutton, 1966). About two decades later Beuret and Webb (1983) found that younger graduate engineers felt isolated from the policy makers of their units, and that the latter, especially those from finance and sales, did not understand technical issues. The relatively low status of engineering in education has been linked to its tendency to shelter under the wing of pure science, unduly theoretical curricula, inadequate recruits, high dropout rates and poor output in terms of the inability and/or reluctance of graduates and others to cope with the often crucial non-technical aspects of engineering.

The lack of status of engineering and engineers in political life has also been noted: Leach (1965) for example criticized liberal Left and humanist intellectuals for 'technophobia', for being 'rampantly negative' and 'wilfully irresponsible' towards engineering. Yet engineers' openness to the outputs of self-styled humanists often put the latter to shame. Both Hinton (1970) and the Finniston report's authors (1980) wanted more national policy and decision makers to be engineers, and the Conservative Political Centre (1978) sought the appointment of a Cabinet Minister 'for engineering'. The latter also wanted the name of the Science Research Council to include the word engineering – as has now happened.

As a sort of preliminary round-up of the types of status-related factor which differentiate most British engineers and foreign counterparts, it is instructive to look at a comparison made between the results of a survey of British mechanical engineers (Gerstl and Hutton, 1966) and a later survey of German mechanical engineers (Hutton and Lawrence, 1981: see Chapter 9 for more details of this comparison). First, the Germans were much better paid than the British engineers, and, compared to other equivalent middle-class occupations in Germany, than the Britons were in Britain. The evidence on British engineers' pay compared with counterparts abroad and with people with similar-level qualifications tends to be rather weak but Hutton and Lawrence's conclusions here are broadly supported by virtually all such evidence (Glover, 1977b; Wilby, 1985; Child *et al.*, 1986). In Germany for example, senior engineers earn more than university professors and senior civil servants: the opposite of the British case. The same international difference applied in the late 1970s to comparisons between private sector engineers (better paid in Germany) and public sector ones (better paid in Britain than private sector counterparts).

Further, German engineering courses are longer and more practical and sector-specific in content than British ones. German engineers are much more likely to be sons of 'professional and executive' families, to have 'middle-class' friends and hobbies, and somewhat more likely to have attended selective secondary schools (grammar or public in Britain). Secondary education in Germany is broader, certainly after the age of 15 or 16, than in England and Wales. British engineers are more likely to think of eventually moving 'into management' – but in Germany, engineering *is*, in manufacturing industry, most of what Britons call 'management'. German engineers are more likely to regard design as central to an ideal engineering course; Britons tend to prefer 'basic engineering science'. Germans are more likely to hold doctorates, to have published books and articles and to have patented inventions. Germans are more likely to work in the private sector, and to work for smaller firms. More Germans had the use of a secretary (72 per cent against 43 per cent). Germans expressed higher levels of satisfaction with their jobs, and were more likely to stay in them for lengthy periods. Finally, Germans worked longer hours. All these conclusions, it must be noted, take the difference of some thirteen years between the dates of the two surveys fully into account, partly with the help of other data.

Differences in status and differences in power and in terms of wealth and class (the last referring to control or lack of control over the process of production) all tend to go together. Compared to equivalent groups British engineers appear to suffer in terms of all these factors. They are kept in their 'specialist' place, not only by non-engineer-dominated managements, but also by such strategies of social closure (which exclude them) and social control (which induce conformity) as 'generalist' management theory and education. If 'management' is – as it certainly is in the Anglo-Saxon countries – seen largely as a matter of finance, marketing, industrial relations and of partly effete social skills, the practical, the technical and the specialist will be readily excluded. Selection for management courses has often been shown to have more to do with 'personal qualities' than with competent or superior technical performance (Whitley *et al.*, 1981). Technical subjects are pretty unimportant in Anglo-Saxon post-experience management education compared with such subjects as marketing, finance and corporate strategy. The Diploma in Management Studies (DMS), traditionally popular with engineers (and much less so with accountants), often includes courses in operations management, but mainly emphasizes the above kinds of subject. So too do the usually much more prestigious Masters degrees in business administration or management: these attract a much higher proportion of accountants and marketing people aiming for top executive posts. They tend to place relatively little emphasis on operations, especially production management, and the (minority of) engineers who gain places on them often appear to see them as a vehicle for moving away from engineering and in particular from production.

One final point concerning the low status of British engineers should be noted. There exists in British popular culture an idea in which an allegedly arcadian rural past is glorified to the detriment of industrial society with its supposed attendant evils. Engineering is implicated in this fall from grace as the perpetrator of the technical changes which led to industrialism. Popular novelists and political pamphleteers like Dickens, Kingsley, Lee, Hardy and others helped to sustain this myth. In point of fact in 1840, when this industrial revolution had supposedly been taking place for decades, only about one in ten of the country's male workforce was employed in a factory, and one which employed 500 people was regarded as huge. Fores (1981) has argued that the fact that the very misleading idea of a dramatic industrial revolution has been taken so seriously is of more significance than the events of the period themselves (see also Kumar, 1978).

Distaste for Manufacturing: a Historical Overview

In order to understand the British distaste for manufacturing the economic and social developments in British society during the eighteenth and nineteenth centuries have to be considered since they provide background information to the issue. The standard studies of British economic development in the decades before 1800 show that much of it took place in farming, commerce, transport and in the growth and distribution of population, as well as in manufacturing. The more significant shifts in employment in what is often taken to be the period of the industrial revolution, between 1760 and 1840, were from the land to services, not from land to manufacturing. Much of the growth that did take place in manufacturing was in the very untypical cotton sector which by 1811 produced about half of Britain's manufacturing output. Cotton never employed more than 7 per cent of the national labour force, and always (in the nineteenth century) employed less than the proportion of the working population employed in domestic services. Farming always produced more income and employed more people than cotton between 1760 and 1840.

By 1840 most workers were craftsmen, labourers or domestic servants. The typical worker was not a machine-tender based in a factory. The domestic or 'putting-out' system of manufacturing was widespread; and so was water power, as opposed to steam. Most British (and European and North American) manufacturing took place in the countryside, near sources of water and coal. Many Victorian 'industrial' cities and towns, like Manchester, Bradford and Hull, developed primarily as centres of commerce and services not manufacturing. There had occurred to 1840 general changes in wealth, in farming, commerce and population, and there certainly was an 'urban evolution' from a rural England to an urban Britain, but even in mid-century, Britain was not an industrial country.

The middle decades – the 1850s, 1860s and 1870s – of Queen Victoria's reign are the key ones for understanding Britain's present situation. 'Modern Britain is essentially a mid-Victorian creation, invented part-consciously to do a specific job which seemed, to certain groups of people, to be the right one at the time' (Fores and Glover, 1975; see also Horne, 1969). That period of 'reform', 'equipoise' or 'improvement', as historians called it, saw the firm establishment of middle-class political power (Briggs, 1959, Burn, 1964, Woodward, 1962). It saw the beginnings of a new central government bureaucracy, which would increasingly be staffed through a new system of competitive examinations. It saw middle-class political, economic and social liberalism come to dominate the thinking of influential people in all walks of life. It witnessed much of the creation of a new kind of English gentleman, more restrained and caring than his predecessors had been, and better able effortlessly to distance himself away from the unseemly demands of hard specialized work.

Both the old landed elite and the new wealthy classes had feared the mob, and both had privileged positions to defend. Rather than fight each other, a process that occurred in some European societies, they built a classic, and for our time the most important, British compromise. The old interests offered a range of political and social concessions. The middle classes were given the vote and non-

Anglicans were allowed to enter Oxford and Cambridge universities. Private education was expanded and reformed to help the sons of provincial factory owners and merchants learn how to deport themselves as gentlemen. From the 1850s onwards Oxford and Cambridge were made more efficient. Established religion in England and Wales became much more serious than it had been in the eighteenth century (partly to help equip itself for debates with natural science). Professional groups were encouraged to develop so as to help make society more efficient and to provide socially respected work for the talented.

In a newly-urban society with a rapidly changing division of labour, a whole range of new leisure-time activities was put together and exported. The basic elements of the British national systems of education and training were established along with an odd habit, one practised much more in Britain than in other industrial countries, of believing that natural science is the ultimate source of wealth. The trade union movement and socialist political activity were largely products of the same period. Many of the laws which still affect modern Britons were passed then (concerning divorce, elementary education, worship, postage and sanitation for example). Friendly societies, qualifying associations for professional men, co-operative societies, two-party 'adversary politics', building societies, democratic local government, popular monarchy, and a belief in the superior mechanical ingenuity and moral qualities of the British (especially the English) over anyone south or east of Calais all owe all, or most of, their origins to this period.

The gentlemanly compromise was celebrated through the development of new lifestyles, as in the 'suburbs' where a cosily idealized rural past lived off the wealth produced in smoky mills and factories. It was a three-class society – upper, middle and working – divided horizontally, whereas in earlier times social divisions had more of a vertical shape, between administrators, the army, the Church, the country, the towns and so on, divisions recognized as different estates of the realm. The Victorian middle classes helped to develop, and then to exploit, religious revivalism to impose their traditional rectitude, even the 'puritanism of the middle ranks' of society, on those above and below them. Middle-class morality and reforming zeal along with beliefs in hard work, self-help and self-discipline, were used to set an example to the working classes, and also to encourage the old upper classes to behave more responsibly than they had in the eighteenth century.

Educated Victorians left their countrymen with two mental legacies which have since been millstones around their necks. One was the over-dramatic perception of the social problems of industrialization, and the other the idea that the processes of manufacturing and technical change are somehow becoming more scientific. Both are best understood as products of a culture disoriented by too much success too easily gained, of a society too ready to believe in its own propaganda.

The British Economic Problem: Symptoms and Explanations

There are a number of commonplaces concerning the British economy: first that is in decline, second that it has declined, and third that Great Britain's rate of economic growth has been and continues to be poor. Although these are ideas which are central to the concerns of this book along with exploring the place of the engineering dimension in this decline, the notion of general economic decline is one that cannot pass without some comment. In the first place the ideas of decline or of slow growth rates imply change of a relative kind. Thus the question is, change compared to what? The conventional measures are ones of decline relative to other economies, particularly to those of other developed countries, and/or compared to Britain's performance at some time in the past. Frequently the commonplace of decline assumes some kind of golden age in the historic past where Britain's performance was better, an assumption which on closer examination needs extensive qualification. In the second place, talking in overall economic terms needs some qualification also. Different economic sectors grow or contract at different rates so notions of decline have to be measured against the performance of various key sectors in the economy. Indeed rates of growth and decline have neither been uniform across sectors nor through time. (For a fuller treatment of these issues and the controversies surrounding them the reader is referred to Aldcroft, 1986; Pollard, 1982, 1983; Kirby, 1981, on which the following account is based; and for an introduction to the theoretical background, Prest and Coppock, 1984).

Overall, the rate of growth of Britain's economy as measured by the Gross Domestic Product (GDP: a measure of the output of the whole economy) increased through the twentieth century until the 1970s when it declined.

Table 3.1 suggests that apart from the most recent period covered by the statistics, rates of growth have been increasing continuously. However that is too simple a conclusion and more detailed considera-

Table 3.1   *Economic Growth: UK 1900–79 (% increase per annum)*

	GDP (average estimate)	GDP per person	Employed labour force	Capital stock (excluding dwellings)
1900–13	1.5	0.6	0.9	1.7
1922–38	2.3	1.2	1.1	1.7
1950–60	2.6	2.2	0.4	2.8
1960–70	2.9	2.6	0.3	4.3
1970–79	2.1	1.9	0.2	3.2

Source: Prest and Coppock, 1984, p. 51.

tion reveals that through the twentieth century there have been many ups and downs. Before 1914, total industrial production was growing and 1913 was the peak year. However while between 1900 and 1913 GDP at factor cost was advancing and output per head was rising at 0.7 per cent both were growing more slowly even then than in the USA and in Germany. British exports also rose during this period at a rate of 0.93 per cent, however, and – this is a crucial point in terms of later economic development – most of this was concentrated in the old staple industries like coal and cotton textiles, a factor that was to have dire consequences later in the century. In spite of this growth, underlying trends were apparent in the British economy before the First World War which have had a profound effect on later development. In coal mining, production rose from 223 million tons in 1900 to 287 million tons in 1913, but the sector was already technically backward relative to Europe and America. British pig iron and steel production also showed considerable growth in the period. However, the increase in the rate of growth in the USA and Germany was faster. Simultaneously Britain's record in newer and more technically advanced sectors like electrical engineering, chemicals, office equipment, machine tools and agricultural machinery was poor, while in older sectors like steam engine construction, textile machinery and heavy machinery the record was good. The chemical and related industries were growing rapidly from a very small base but they were often technically and scientifically deficient compared with Germany and the United States. For the workforce, and in spite of the overall economic growth, real wages were generally static and widespread differentials between skilled and unskilled workers led to gross inequalities (Pollard, 1983: 1–19). Thus the structure of the British economy with its concentration on older industries, technical backwardness, and slow growth relative to some others, was established before the First World War.

The First World War created special problems for the British economy, a fact which was only gradually realized by politicians as the war progressed. The diversion of huge amounts of resources into war production and to the armed forces played havoc with market forces, and as the war progressed the state took an increasingly interventionist stance in economic affairs. However, total output of British industry hardly declined at all, and some commentators believe that it even increased slightly. Some new industries emerged, and some, along with some older ones, grew rapidly, for example scientific instruments, ball bearings, specialized glassware and certain chemicals; but Britain was left with a large surplus capacity in shipbuilding at the end of the war (Pollard, 1982: 26–30).

In the decade after the First World War economic growth was slow and patchy and by 1929 Britain had high levels of unemployment, stagnating exports and a massive regional imbalance in terms of declining industries in the 'older' industrial areas. At the end of the war, there had been much pent-up demand in the economy. People wanted goods and services, but the supply of such goods and services was unavailable. A result was price rises on a large scale. The accompanying boom and associated speculative frenzy lasted only to 1920 and in 1921 there followed what Aldcroft has called the worst depression in Britain's history – there were some 2.4 million people unemployed by May 1921 and on all major economic indices Britain's economy was in decline. Sadly little use was made of the experience which had been gained in wartime in improved production techniques and working practices and the economic controls which had been imposed by Government had been quickly undone. The customs and practices of pre-war days returned to British industry and while productivity did not change, costs, particularly labour costs, rose considerably. There was a short recovery in the mid-1920s but its amplitude was small and it came to an abrupt end in 1926 following the General Strike (Aldcroft, 1986: 1–41).

The later 1920s saw a reversal of fortunes in Britain compared to the United States, because Britain experienced a 'feeble' boom. However, relatively speaking, Britain continued to sink down the economic league table. But the situation varied from sector to sector. Unemployment was concentrated in mining, mechanical engineering, shipbuilding and iron and steel and textiles and in mid-1929 these sectors accounted for half of the unemployed population who were part of the Government's insurance scheme. These so-called staple industries suffered a particular collapse in this decade, not only because of their technical backwardness, but because of their reliance on exports. The demand for their products overseas was drying up and the regions in which these industries were located, South Wales, West Scotland, North-East England and Lancashire, paid a heavy price. Meanwhile the newer industries, whose products were geared to the domestic market, were unable to fill the output gap. For the most part the newer industries, like motor manufacture, electrical engineering, precision instruments and chemicals were located away from the declining regions. In Aldcroft's view the whole matter was exacerbated by the macro-economic policies pursued by governments which were predominantly concerned with maintaining the currency, but which were at odds with the needs of the domestic economy. During the 1920s British exports remained stagnant while the volume of world trade rose by 27 per cent. Britain's export capacity located in its staple industries simply did not respond to the changing pattern of world trade. Consequently Britain's share of world exports fell from 29.9 per cent to 23.6 per cent (Aldcroft, 1986).

Etched on the British collective consciousness is the idea of a terrible depression during the 1930s. With for example an estimated 3,289,000 people unemployed by 1931 rising to 3,750,000 in September 1932, and bearing in mind the fact that a much higher proportion of these were their families' only breadwinners than is the case with today's unemployment, there appears to be ample evidence to support the commonly held view. But once again it is a view that has to be qualified. For example, from 1914 to 1938 the standard of living for the total population did rise, but not for all the population, only for those in regular employment. In York, for example, in 1936 an estimated 31.1 per cent of the working-class population was living below the poverty line. Yet the scale of deprivation nationally was not as bad as those of Britain's neighbours and competitors. Even though, in growth terms, through the 1930s Britain fared better than most, there were major problems, and trends evident in the 1920s and earlier continued. In 1929–32 the volume of exports fell by 37.5 per cent. In 1931 there was a balance of payments deficit of £114 million. Decline in real terms was 5 per cent or less. Industrial productivity fell by 11.4 per cent (1929–32).

There was an international financial crisis in 1931. However, in Britain a recovery began relatively quickly – but it was an economic recovery which was domestically based, that is to say it was based on domestic demand, not on Britain's exporting capacity. The first full year of British recovery was in 1933 and even then the volume of exports remained stable. The real thrust for recovery came from the non-export sectors, building, road transport, electricity supply, electrical engineering and vehicles and consumer durables. In 1934–5 exports (except coal) also picked up. There was something of a boom through to 1937 followed by a recession in 1938, but even then industrial production and domestic output declined only 1 or 2 per cent, while real incomes and consumption continued to rise. During the 'boom' years from 1932 to 1937 real incomes and consumption continued to rise, real income by 19 per cent, GDP by 23 per cent, industrial production by 46 per cent, investment by 47 per cent. However, regional unemployment located in the declining older sectors continued to blight the country. Aldcroft argues, though, that during this period Britain's performance was better than that of Germany, the United States, France, Belgium and Canada – it is however the only period in the twentieth century when Britain did outpace her main competitors. And then Britain found herself embroiled in another war (Aldcroft, 1986: 44–61).

The war decade, the 1940s, was one of austerity: shortages and rationing were among the dominant themes, but it was a time nevertheless of full employment and of greater equality in income distribution and consumption, as well as of massive state intervention in the economy. During the war defence spending accounted for a very large chunk of resources. In 1939 defence spending was 21.4 per cent of Gross National Product (GNP). By 1940 and for the rest of the war Britain's defence spending accounted for over half of the national income. The structure of employment also changed; there was a large shift of manpower into the metal trades and engineering, shipbuilding and chemicals, and by 1942 unemployment was down to 0.5 per cent. The state accumulated immense powers over production, consumption, investment, distribution, transport, building, foreign exchange, imports, manpower, prices, raw materials and the nature of many goods. Aircraft production, electrical and general engineering, metal working and chemicals experienced the greatest physical increase in activity, while overall output per worker rose by 10 per cent between 1938 and 1943.

Agriculture became one of the most dynamic sectors. Agricultural production rose by 35 per cent, while output per man year rose between 10 and 15 per cent, so that Britain came to have (and still has) the most efficient agricultural operation in the world. Through the war years coal was the least successful of Britain's industries. However, the problem which was being stored up by the tremendous volume of economic activity at home, by fighting activity overseas, as well as by American pressure in return for money to finance the war, was that British export markets abroad were neglected. The war had also massively drained the country's financial reserves and its economic strength in general.

At the end of the war, significant problems arose. Apart from three acute financial crises with which the government had to contend, the home economic infrastructure was badly depleted. Most seriously Britain had been transformed from being the world's largest creditor to the world's largest debtor nation. Capital stock had deteriorated. But there were some bright spots which offered potential. Engineering and metal working had been partly transformed for the better with the wholesale use of mass production techniques and manufacturing industry had not suffered any long-term damage to its efficiency. The war had boosted growth in the more technically advanced sectors and had pulled the country away from its dependence upon the older industries (for fascinating and challenging accounts of these issues and of underlying problems of poor levels of skill and anti-industrial values, see Barnett, 1972 and 1986).

The Labour Government elected in 1945 continued with many of the economic controls which had come into force during the war and planned for full employment. Output per head continued to rise and between 1938 and 1950 manufacturing industry expanded faster than other sectors with the greatest increases to be found in metals, engineering, chemicals and related industries, and machine tools. In the years from 1945 to 1951 Britain's economic performance was quite impressive, but from the vantage point of 40 years afterwards it was also a time of massive missed opportunity. On all major economic indices there were percentage increases (GDP rose by 12.1 per cent, industrial production 32.1 per cent, manufacturing 36.8 per cent, exports 73.1 per cent, total consumer expenditure 7.5 per cent and per capita expenditure 5.8 per cent). The great post-war upsurge in production really gained momentum in 1947. Those industries which had done well in the war continued to do so afterwards – all of them engineering-based – manufacturing, gas, electricity, water, transport and communications, and these industries in the main account for the growth in the United Kingdom's global productivity in this period. In 1944 exports had been 38 per cent of their 1938 level, but by 1946 they had returned to their pre-war level and by 1951 they were up 75 per cent. Moreover the product mix in exports was now weighted towards the new and expanding industries rather than towards the old staple ones. So why then did this pattern of growth not continue?

Well, at one level growth did continue until relatively recently. GDP grew continuously from 1950 to 1979 and in the same period there was considerable change in the industrial structure as the old staples declined sharply and the newer industries expanded. Other industries showed economic growth but also because of changing techniques they shed labour and streamlined. The car industry grew rapidly until the 1960s when import penetration began seriously to erode domestic producers' market share. The chemicals sector grew rapidly although it did not capitalize on its wartime successes, and it was generally slow to innovate, particularly in plastics. Agricultural production and productivity expanded enormously. Into the 1970s oil was the fastest growing sector in energy and appeared to be at the forefront of technical and engineering knowhow. But the key to understanding the British situation here lies in understanding not that Britain has experienced growth, but rather that the rate of growth in postwar Britain has been much slower than most of her competitors for most of the time. Indeed Pollard argues that if Britain had experienced 'a rate of economic growth similar to that of the rest of the advanced world [it] would have almost doubled the standard of living [of] that obtained in the UK in the 1970s and would put Britain at the level of the richest countries in Europe' (Pollard, 1983: 347).

The turnaround in Britain's position is demonstrated broadly in Tables 3.2, 3.3 and 3.4.

Because of the vagaries of such broad statistical comparisons as these – categories vary between countries, exchange rates alter in

Table 3.2   *Annual Growth Rates of Real Gross Domestic Product (Selected Countries) 1955–73 (% per annum)*

	1955–60	1960–4	1964–9	1969–73
UK	2.5	3.1	2.5	3.0
France	4.8	6.0	5.9	6.1
West Germany	6.4	5.1	4.6	4.5
Italy	5.4	5.5	5.6	4.1

Source: Kirby, 1981, p. 146.

Table 3.3   *Growth of Industrial Production: UK and Selected Countries 1960–82. Annual average increase in industrial production (%)*

	1960–73	1973–5	1975–9	1979–82	1960–82
UK	2.9	−3.8	3.6	−2.7	1.7
USA	4.9	−4.7	5.7	−2.9	3.2
Japan	12.6	−7.1	9.2	1.5	8.2
France	5.9	−2.4	4.2	−0.8	3.9
West Germany	5.5	−3.8	4.0	−1.2	3.4

Source: Prest and Coppock, 1984, p. 190.

Table 3.4   *National Income per Head: UK and Selected Countries, 1960,*
1973 and 1982 (incomes per head in pounds sterling at prices and
exchange rates relevant each year)

	1960	1973	1982
USA	818	2304	7243
Canada	578	2027	5992
Japan	135	1331	4315
France	368	1745	4639
West Germany	364	2031	5909
Italy	203	955	3391
UK	398	1409	4276

Source: Prest and Coppock, 1984, p. 193.

relation to each other all the time, and the composition of incomes
varies (for example it costs a lot less to keep warm in Italy than in
Canada) – too much should not be read into the details of the above
figures. However they can be taken as being broadly representative of
the major international differences over the periods which they cover.
One caveat is worth making, however. It is now (in 1986) generally
accepted that the Japanese and West Germans are as well off as the
residents of the United States (and the relative position of Britain has
continued to worsen since 1982).

In general terms, then, Britain has experienced a process of relative
economic decline. The question arises as to its cause. Sorge (1978a)
has argued that most explanations of British decline only described
parts, symptoms, or effects of it. Most debates had been 'too superficial and unconcerned with what went on where wealth was created'.
Sorge referred to the following conventional explanations of British
decline: underinvestment, the growth of the public sector, high
taxation, over-regulation of industry and employment, industrial
disputes, low productivity, the low status of work in manufacturing
industry and the neglect of manufacturing and engineering. Taking
each of these ideas in turn we find (following Sorge) all bar one to be at
least partly inadequate.

First, more has often been invested in manufacturing in Britain than
elsewhere. The real problems are the low rate of return on *existing*
capital, the high proportion of investment in military and glamour
projects (like Concorde) and financial institutions' readiness to allow
capital to be exported.

The notion of the manufacturing firm as a system for converting
material and human inputs into saleable products entails a medium

term plan for the processing of resources, but finance in Britain has traditionally emphasized the balance sheet and short term profitability. Firms which have invested in the future have made themselves liable to takeover, usually by larger more powerful firms, as their cash value rises with liabilities for the future being discarded as their prospects for lengthy survival increase. Comparisons with France, Germany, Japan and the United States have been used to highlight differences in ways in which industry is financed. The relationships of the banks with industry in all of the other countries have been closer than in Britain, although the British and American approaches have had more in common than the others. However relations between industry and finance have become much closer in Britain than they were in the first half of the 1970s and earlier, with large companies having become more reliant on bank borrowing and expanded medium-term loan facilities. The change has been brought about by the depth of recession, because financial institutions have been forced to play a more active role, helping companies in difficulties to re-structure themselves. Also institutional investors like pension funds have been taking a closer interest in the affairs of the companies in which they have invested (Vittas, 1986). There are therefore some grounds for thinking that finance and industry are working more closely in Britain, and for feeling more positive about the role of finance in Britain than has been possible for a long time. Even so there is little doubt that strategic thinking in much of British industry has long been hampered by neglect and even hostility towards its needs on the part of the financial sector and that such people as accountants, bankers and engineers still have a lot to learn from each other. Important as this is, however, it is only part of a more fundamental problem.

Sorge, on the second 'conventional explanation' noted that the growth of the public sector has been less pronounced in Britain than in several economically more successful countries. However the character of British public sector management may be a problem, being over-bureaucratic and with many job holders inadequately qualified. The data on high taxation are equivocal, Really high levels of taxation have only existed for short periods in Britain. In any case taxation has long been much higher in Sweden than in the UK, yet (and in spite of some recent problems) the Swedish economy is a strong candidate for the postwar success story of Europe. The over-regulation of industry and employment in fact consists mainly of Labour legislation (of the mid-1970s and earlier) affecting hiring and firing procedures. The fact is that more successful and more genuinely capitalist countries like West Germany had long had much more in the way of legal controls over employment and training than exist even in the 1980s in Britain (Sorge and Warner, 1986).

Next, there is the perennial question of Britain's poor record in industrial relations. Britain has been too low in the international league table of worker days lost to suggest that this is a particularly 'British sickness'. Strikes appear to be more of a symptom, part, or effect of the economic problem than anything else, even if the relatively large size of British organizations means that the 'knock-on' effects of disputes can sometimes be considerable. Low productivity and the low status of work in manufacturing industry and poorly qualified management are all true, but are again symptoms, not the problem itself. Finally, the neglect of manufacturing and engineering we define as the basic problem. And this we expand and elaborate on both in this chapter and the rest of the book.

It is interesting to speculate a little about the party political and other directions from which have come the explanations which like Sorge we criticize. Those who argue about underinvestment tend to come from the political Left, from trade unionists, Labour and sometimes Alliance politicians, and from Left-minded academics. The argument sometimes regards British decline as a special case of the decline or 'contradictions' of capitalism in general. The argument can appear valid in the British case since Britain's economy remains a predominantly capitalist one. However it is undermined by the fact that several other major capitalist countries have not suffered in the same way as Britain. Nevertheless there does appear to be some truth in the related argument that Britain has tended, both traditionally and in recent years, to export capital rather than to invest it in domestic manufacturing industry and that this is a typical feature of British 'imperialism' in particular and that of 'capitalism' worldwide (cf. Pollard, 1982; Frank, 1980; Ingham, 1984). Thus although lack of investment in manufacturing is not a cause of Britain's problem, both the use of existing capital and traditional British habits concerning the direction of investment are important symptoms and parts of it.

The political Right has tended to favour arguments about the growth of the public sector, about high levels of taxation, and about state regulation of industry. In each case the details have been more complex than our rebuttals suggest. For example, and while its extent and significance have often been exaggerated, the growth of the public sector of employment occurred for many diverse reasons. These have included increased social security spending in the recession, technically more sophisticated health care, more civilian back-up for defence activities, and state takeovers of large parts of ailing private industries such as motor manufacture, shipbuilding, and steel. Since

1979 and the election of governments which have slimmed down the civil service and privatized state-owned organizations and industries the argument about the public sector's growth has become rather dated.

It is true that very highly paid people in Britain have been more highly taxed than many counterparts elsewhere. It is hardly surprising that those who favour the high taxation thesis have been influential and relatively wealthy people (cf. Edwardes, 1978). The regulation of industry argument seems to have been particularly popular amongst spokesmen for industry and managers such as the Confederation of British Industry and the Institute of Directors. Similar pressure groups have focused on employment (including health and safety at work) legislation as well as the collection of Value Added Tax as a hindrance to dynamism and profitability in the British economy. As suggested previously such legislation had long been established in other capitalist countries without noticeable ill effects before they impinged on British employers.

The broadly right-wing idea that strikes are the 'British sickness' is probably not as popular as it was a decade or so ago. At one level more people have since had the experience of belonging to trade unions and of hearing other views. More important, however, has been the worsening underlying economic situation, which has meant fewer disputes while provoking more serious study of the roots of economic decline, and highlighting the superficiality of immediate, sensational kinds of explanation. We do not however want to deny the fact that English culture's emphasis on individual freedom has traditionally been lenient towards some of the narrowly self-serving behaviour of some British trade unions (Mant, 1977; Macfarlane, 1978; Maitland, 1980; Marengo, 1979; Olson, 1982).

When the 'low status of manufacturing' is associated with 'contempt for wealth creation' speakers tend to come from the political Left. Yet given the scale of the Labour Party's 1983 General Election defeat, Labour (and SDP/Liberal thinking) has increasingly begun to focus seriously on wealth creation, rather than on its distribution.

It has been suggested that growth in government spending has pre-empted resources available for investment in manufacturing and jobs in it (Bacon and Eltis, 1976). Yet there was strong growth in manufacturing investment throughout the 1960s, and greater public spending has probably reduced Britain's ability to import competing foreign goods (Cairncross, 1978). Britain's increasing failure to export and to prevent import penetration imply a vicious circle of declining competitiveness, markets, innovation and investment. Britain's balance of trade in manufacturing has steadily deteriorated.

The base of potentially exportable services was too small to make up for this and productivity gains in services rarely equalled those in manufacturing. North Sea oil's benefits to the balance of payments have only been helping to cover up the basic problems in manufacturing and elsewhere.

'Deindustrialization' is not due to technical development as various economists, 'science' journalists, sociologists, trade union leaders and politicians have sometimes argued. Technical development does generally raise output (and incomes), and displace labour, in the short run. However its overall effect is to increase wealth which is then available for investment in new activities and jobs. The introduction of sophisticated processes (so-called high technology) normally involves the use of a great deal of, mainly highly skilled, labour so the question of whether or not it displaces labour in the short term depends on whether or not the new processes are imported. Thereafter the issue is simply one of building on success and both common sense and basic economics would suggest that a successful innovative organization is likely to encourage more employment, direct and indirect, than a struggling non-innovative one. The argument that sees technical development as a major threat to employment also fails to explain why countries at the forefront of technical development generally lose fewer (or gain more) manufacturing and other jobs than the laggard countries. Lawrence (1983) argued that it made more sense to link Britain's gradual relegation to Europe's technical and economic 'periphery', making Britain more comparable with Eire, Italy or Greece, than with France, Sweden or West Germany, to Britain's growing relative technical backwardness, not to technical development as such.

Contributions from Sociology

Twentieth-century sociology's tendency to neglect engineers probably owes a great deal to an important change in the European intellectual climate at the end of the nineteenth century (Hughes, 1968). During that century scientific study of such phenomena as evolution and electricity had helped to popularize a rather mechanistic set of ideas about human nature and relationships. To many, change, desirable or otherwise, had an inevitable, almost automatic, quality. Late eighteenth- and early nineteenth-century economic and political thinkers had optimistically sought rational solutions to social problems. By the late nineteenth century natural science seemed, to many intellectuals, to have won its battles with religious faith. All this conspired to produce a fatalistic, mechanistic, even claustrophobic, belief in material progress. Many felt that such progress would come inevitably, either through the survival of the fittest under capitalism or through class conflict and revolution.

Several of the founding fathers of social science opposed this extreme and pessimistic kind of outlook in the decades around 1900. To replace the reigning cult of mechanistic positivism, Freud in psychology and Durkheim and Weber in sociology offered thoughts about human nature and social science which respectively stressed the importance of emotional aspects of motivation and of social bonds, and the desirability of free speculation and rational inquiry. Beneath their powerful scientific curiosity, their thought was also affected by the unstable and to some extent decaying nature of the European societies in which they lived. This made their concerns rather general and often more to do with ways of looking at the socio-psychological world, than with the world itself. Like Marx, Durkheim and Weber were far more interested in relationships at work than with work as such. So although Weber, and especially Marx, depicted man as *homo faber*, the maker and doer, for which the engineer provides the outstanding example, a great deal of twentieth-century social science has tended to play down the ways in which individuals continually act to confound its generalizations about social relationships.

In spite of this general tendency to neglect the details of economic life, a number of sociologists have produced valuable studies of engineering and related matters since the Second World War. These include Cotgrove (1958) on higher technical education, Burns and Stalker (1961) on the organization of electronics firms, Gerstl and Hutton (1966) on the backgrounds, careers and social and other attitudes of mechanical engineers, Lawrence (1980) and Hutton and Lawrence (1981) on West German (contrasted with British) management and engineering, Fidler (1981) on the company directors whose actions and attitudes often determine what engineering will consist of, Beuret and Webb (1983) on younger engineers' experiences of work and their strengths and weaknesses as potential executives, Ingham (1984) on longstanding divisions between the worlds of industry and high finance, Sorge and Warner (1986) on factory organization and management in Britain and Germany, and Whalley (1986) on the use of engineering knowledge and the position of British engineers in the class structure. All of these studies contain a great deal of potentially useful evidence and ideas, and all are largely jargon-free, intelligible and interesting to laymen. Several of the studies have influenced government policies for engineering education and training. More radical sociologists than these have also produced studies and arguments critical of the roles of engineers and of those who manage them (see, for example, Collingridge, 1981, and Hales, 1982). Although some engineers may not always enjoy reading criticisms of what they or their colleagues do the material is usually both quite interesting from a technical standpoint and thought-provoking from a socio–political one. Notwithstanding this, however, sociology normally offers little in the way of solutions, at least directly.

The critical issue which sociologists and indeed other commentators have failed to acknowledge is the importance of industry as a system – a system in which engineers play a central and vital part. The system includes design, development, production and marketing. The links between these things are fundamental to the success of the system. British industry, British technical, business and management education and British social scientists have all tended to treat these aspects as discrete, partly separate functional elements, which in one sense of course they are. But, the division of labour, which these different activities manifest, is a holistic and a systemic idea. So not only has British industry, unlike say Japanese industry, failed to grasp the holistic nature of the enterprise with the market pulling through all other activities, commentators on the topic have also traditionally failed to appreciate the problem (and most still do). The notion of the division of labour comes closest to articulating the idea of system in a sociological sense, because it emphasizes not just divisions (separate functions) but also the end product – the purpose for which labour is divided.

The status issue of the engineering role here is writ large. It is precisely because of the fragmented nature of understanding of the system, that 'common sense' allows for one or more parts to be denigrated at the expense of others, when in fact all the parts need each other for the system to function at all. Underlying this problem is a complex of attitudes towards industry, education, occupations and the use of knowledge for practical ends, not to mention the content and organization of education and training, of occupations and professions, and the ways in which employment is typically organized. The attitudes both reflect and help to produce the problems with the division of labour with which we are concerned.

The Searches for Solutions

Concern with British inefficiency in manufacturing was voiced as early as the 1830s, well before Britain had become a recognizably industrial country (Barnett, 1972). But it was not until the 1870s that British governments began to be seriously worried about foreign competition and weaknesses in manufacturing at home. Since then there have been many public inquiries into engineering and scientific education and the status and roles of engineers and scientists, into industrial relations, management efficiency, management education and into the provision of finance for industry. Thus Finniston (1980: Preface) commented on how many previous inquiries into various aspects of engineering and engineering education there had been, and on how their recommendations had usually been ignored or otherwise ineffectual. Having read several of them, including some of the very earliest, we are inclined to agree with Finniston that many of their findings remain 'valid still, even . . . many years since their issue'.

Inquiries have been conducted by Baron Lyon Playfair (1852, 1853), the House of Commons Select Committee on Scientific Instruction under Samuelson (1867–8) followed by a full-scale Royal Commission under the Duke of Devonshire (1871–5), the Institution of Civil Engineers in 1870, another Royal Commission on technical education (1884), the Civil Engineers again in 1906, the Balfour Committee on Industry and Trade (1929) and numerous others, notably Fielden (1963) on engineering design, Jones (1967) on the 'brain drain' or emigration (mainly to the USA) of engineers and scientists, Dainton (1968) on the flow of students into higher engineering education, and Swann (1968) on their subsequent passage into employment. There was a plethora of reports in the late 1970s and early to mid-1980s from the Departments of Industry, of Education and Science, and of Employment, from the British Association for the Advancement of Science, the British Institute of Management, the National Economic Development Council, the Institute of Personnel Management, the Manpower Services Commission, the Central Policy Review Staff, the Schools Council, the Industrial Society, the Science Research Council, the Confederation of British Industry, the University Grants Committee, the Science Policy Research Unit at Sussex University, the Engineering Professors' Conference, the Goals of Engineering Education Project at Leicester Polytechnic, Southampton and Loughborough Universities (the work of S. P. Hutton and P. A. Lawrence), Bradford University (S. H. Wearne and others), and The City University (Herriot and others, Levy and others). These were variously concerned with the education, backgrounds and careers of British engineers and managers compared with foreign counterparts, the nature of higher engineering education, most other types and level of education and training for industry, the links between innovation, production and markets, engineers' needs for 'managerial' elements in their education, employers' perceptions of the qualities of engineering graduates, and the preference of many of the latter for jobs in research and development rather than in production.

Politicians have shown a lot of concern, especially since the Second World War, with employment relationships. This however has been symptomatic of an arm's length attitude to economic life. Thus, instead of concentrating on the detailed requirements of tasks to be done and of how to produce people to do them with suitable abilities, politicians have been much more interested in relationships between existing job holders. In the 1960s and 1970s governments first modified and then began to abandon a century-old tradition of governmental, administrative and legislative non-involvement in employment relationships. The factors responsible included the nation's relatively poor economic performance, the end of an imperial role whereby once-exported tensions came home to roost, inappropriate managerial attitudes and abilities, a rediscovery of overmanning and associated restrictive practices on the part of trade unionists, and growth of public sector employment in which the state was directly involved. Since the 1960s the law has been applied, if not often forcefully or effectively, to prices and incomes, the closed shop, redundancy, the rights to belong to trade unions and of trade unions to be recognized by employers, unfair dismissal, training, health and safety, sexual and racial discrimination in employment, the disclosure of information to employees and their representatives, and many other matters. By the mid-1970s a 'basic floor of (legal) rights in employment' was apparently established (Hawkins, 1978), bringing Britain in line with several other Western European countries.

One of the intentions behind this legislation was to curb inflation and abuses of trade union and managerial power, to help free the labour market and to produce more and more effective use of skills. Since 1979 government policy has had similar aims but it has been less sympathetic to organized labour and it has operated in a harsher economic and social climate. It has also sought to reduce the state's direct involvement in employment matters by curbing some of the rights previously granted and by relying on a partial return to reliance on market forces to regulate employment.

The employment legislation of the period from the early/mid-1960s to the mid-1970s culminated largely in the Health and Safety at Work Act (1974), and especially in the Trade Union and Labour Relations Act (1974) and the Employment Protection Act (1975). It was perhaps best understood in retrospect as a kind of housekeeping exercise which might clear the decks for action on more fundamental issues such as productivity, education and training, management quality, industrial democracy and attitudes to occupations and work. Unfortunately the important Royal (Donovan) Commission on Industrial Relations (1965–8), which influenced a great deal of the legislation, tended to assume that private manufacturing industry should be the main focus in attempts to reform industrial relations. However the public and private service sectors were becoming more important not only as those in which far more people worked than in manufacturing but as sectors in which a higher proportion of conflict occurred.

The nationalized industries (and the health and social services), many established after the Second World War, were conceived as much as social services to atone for the hard times of employees over previous decades, as industries. In consequence, and particularly for example in coal and iron and steel, opportunities for increasing efficiency were not seized. Nor was there enough effort put into generating new industries to absorb those who were bound to be made redundant by the older ones (Barnett, 1986). In the 1950s and 1960s there was poor management of get-rich-quick industries with too much emphasis on short-term profits and wage rises, and too much political manipulation of the nationalized ones and also of the important motor industry much of which eventually had to be nationalized (Dunnett, 1980). Private-sector engineering increasingly suffered from foreign competition and workers there became less and less militant in the 1970s.

With decreasing justification the shop floor was thought of as the cockpit of conflict, whereas the varied employment problems of growing numbers of white-collar and so-called knowledge workers were neglected (cf. Kelly, 1980, 1983; Kelly, Martin and Pemble, 1984; Prandy, Stewart and Blackburn, 1983; Hyman and Price, 1983). In the early/mid-1980s the increasingly restrictive character of employment law and of managerial control were helping to qualify severely the optimism which had surrounded the changes of the 1960s and 1970s (Taylor, 1982; Wedderburn, Lewis and Clark, 1983; see also Nicholls, 1986).

The recession has weakened the hand of the trade unions and strengthened those of employers and of government: in the short run it has probably helped to improve productivity through a 'shake-out' of underemployed people. However neither British nor foreign evidence suggests that aggressively militant trade unionism has ever been much more than an interesting and spectacular detail of Britain's problems. Strong trade unions which co-operate with and are effectively co-opted by employers are a different matter, however, if

Swedish and West German experiences are to be believed. Unfortunately traditional British individualism does not seem to favour
systematic long-term co-operation between unions and management. The 'Dunkirk Spirit' of everyone pulling together in
emergencies seems to be reserved for emergencies.

Conclusion

This chapter has been concerned both to contextualize the problem
which was delimited in Chapter 2 and to develop the argument
further. We argued in the previous chapter that engineering could be
thought of as having a crucial role in social and economic development, if a view of human nature, called *homo faber*, was adopted. In the
present chapter the evidence concerning the status position of
engineers in Britain was used to demonstrate the fact that in spite of
Britain being an industrial society its engineers are not accorded a
particularly high status position and that there were powerful forces
which tended to deny the centrality of the engineer's position. We
located the origins of the low status within the kinds of social
institutions and traditions established during the period when Britain
was becoming an industrial society and in particular the social
structure and its associated values which were established in and
around the years 1850 to 1870. Moving the account forward we then
considered data concerning Britain's relative decline as an industrial
country. Critically we argued that for the most part those authors and
commentators who have written about the processes of decline had
misconceived the problem and in general concentrated on symptoms
rather than causes. In other words the importance of productive
activity in an industrial economy did not tend to be the starting point
for their analyses. We concluded that social scientific findings
pertinent to the problem appeared to have been largely ignored by
those in power.

4 The British Economy in Context

Introduction

In the previous chapter the relationships between the low status of engineering, distaste for manufacturing industry and under-performance of the British economy were outlined. But the question of underperformance begs a further question, that of under-performance relative to what? Another issue is raised by the previous discussion. Major national economies do not exist in a vacuum. The British economy is part of a world economic system: this introduction was originally written using a pen made in the USA, on paper manufactured in Scandinavia, on a table made in Britain from African timber, while the writer was drinking coffee containing Scottish milk, West Indian sugar, and coffee beans from Brazil imported by an American multinational, in a cup made in France, and in a room constructed from Scandinavian timber, English bricks and cement and Scottish sand. Although this point is a little glib it nevertheless illustrates the now very considerable interdependence of much of the world's economy. Such interdependence means that the economic performance of nations cannot be studied effectively without reference to world influences and trends. This chapter provides a brief review of some of the important influences on Britain arising out of its economic interdependence. The decline of Britain relatively in economic terms within this interdependent system is quite well known and has been discussed by many commentators. Here we summarize some of the facts and viewpoints.

The Major Industrial Economies

Table 4.1 shows selected economic indicators for several major industrial countries.

Japan has been the star performer among the major industrial countries since 1945 and Britain has been the worst. The USA's

Engineers in Britain

economy has grown slowly but from such a strong position that inhabitants of other countries have only recently begun to enjoy similar incomes. West Germany's performance has been very good, and France's good.

Relative to land area, population and natural resources, Britain underperforms compared to other advanced industrial countries.

Table 4.1 *Selected Economic Indicators for Several Major Industrial Countries*

	France	Italy	Japan	UK	USA	West Germany
Land area (thousand km^2)	547	301	372	244	9,363	249
Population (millions)	53.5	56.9	115.9	55.8	220.6	61.3
Gross Domestic Product (billions US$ at market prices)	471.6	218.3	973.9	309.2	2,112.4	638.9
Share in world's Gross National Product (%)	4.9	2.4	10.0	3.2	21.8	6.6
GDP per capita (US$)	8,850	4,590	8,480	5,530	9,660	10,420
Gross fixed capital formation (% of GDP)	21.5	18.8	30.2	18.1	18.1	21.5
Gross fixed capital formation, transport, machinery and equipment (% of GDP)	9.1	7.8	10.9	9.2	7.3	8.9
Merchandise exports (billion US$)	98.0	72.1	103.5	90.8	181.8	171.5
Merchandise imports (billion US$)	106.9	77.8	110.7	102.8	207.1	157.8

Sources: World Bank, *World Development Report, 1980*; OECD, *OECD Economic Surveys: Japan (1980)*.

The Japanese saved and invested harder than the others, and the average West German was apparently about twice as well off as the average Briton. As we noted in the previous chapter some more recent estimates than those in the tables suggest that the typical Japanese was earning as much as the typical American by the early 1980s. For the same countries in Table 4.1 the 'robot populations' (simply the number of robots installed) in 1983 were: France, 1,500; Italy, 1,800; Japan, 16,500; Britain, 1,750; the USA, 8,000; and West Germany, 4,800. Here the 'advanced' countries are West Germany and (especially) Japan; Sweden with a population of only 8 million and 1,900 robots was also very well favoured (NIER, 1984). The British figure is ostensibly a fairly respectable one, but only about a quarter of the robots in Britain were British-made or assembled. Microelectronics are central to the use of robots as well as to other advanced processes like flexible machining systems and the use of computers in production control. However Britain is not a prominent maker of the 'chip', the basic component of microelectronics. There are exceptional British companies which are pioneers in specialist parts of the market for microelectronics components but in 1983 none of the ten top suppliers to the European market for integrated circuits was British.

United Nations data on 'industrial strengths' in 1980 are the subject of Table 4.2.

Table 4.2   *Indicators of Industrial Strength, 1979*

	France	Italy	Japan	UK	USA	West Germany
Coal, lignite and brown coal (million metric tons)	21.1	1.1	17.64	122.8	38.0	130.6
Natural gas (billion K cal/m³)	70.6	125.4	25.9	357.8	4,846.1	173.5
Crude petroleum (million metric tons)	1.2	1.7	0.5	77.8	420.5	4.8
Crude steel (million metric tons)	23.4	24.4	111.7	21.5	124.3	46.0
Aluminium (million metric tons)	0.6	0.3	1.8	0.5	5.7	0.4
Cement (million metric tons)	28.8	39.7	87.8	16.1	70.5	35.5
Passenger cars (millions)	3.7	1.5	6.2	1.1	8.1	3.9
Commercial vehicles (millions)	0.5	0.2	3.5	0.4	3.0	0.3

Source: United Nations, *Monthly Bulletin of Statistics*, XXXIV, 7, July 1980.

For its size, Britain was relatively strong on energy output, and relatively weak on steel, motors and cement (used for construction) for a

supposedly mature industrial country. Japan, another offshore island
economy like Britain, appears remarkably successful at making and
building things, despite her dearth of natural resources. With a much
larger land area, population and natural resources, the USA does not
look particularly successful compared with Japan and West Germany.

Current Concerns

The last points are broadly reinforced by comparing the figures for
growth of Gross Domestic Product in the period 1972–82 in the same
countries. The figures are: France, 2.7%; Italy, 2.6%; Japan, 4.3%;
the UK, 1.4%; the USA, 2.1%; and West Germany, 2.0% (NIER, 1985).
These figures cover the periods of both the 'oil shocks' or 'energy
crises' of the years 1973–5 and 1979–82, and they are generally lower than
1950s, 1960s and early 1970s figures for growth of the same economies.

Allen (1979) presented some statistics on mid-1970s trends for most
of the same economies:

Table 4.3   *Comparative Economic Performance, 1970–7 (1970 = 100)*

	Gross National Product		Industrial production		Output per man-hour	
	1973	*1977*	*1973*	*1977*	*1973*	*1977*
UK	111	111	111	106	118	121
USA	118	124	120	127	113	127
France	113	131	120	126	121	139
Japan	129	145	127	127	133	155
West Germany	112	119	113	116	117	139

Source: Allen (1979:78).

These data suggest that the energy crisis of the mid-1970s had a
particularly bad effect on Britain. But the energy crises of the mid-
1970s and the early 1980s only served to highlight Britain's long-term
problems. For example Gomulka (1979) showed how the average
American worker produced twice as much in 1974 as in 1950, the
average West German three times, the average Japanese nearly eight
times, the average Frenchman nearly four, the average Italian nearly
three, the average Russian over four, the average East German
between four and five, the average Pole or Czech nearly four, and the
average Briton less than twice as much. Maddison (1979) described
how the USA's lead in productivity over other Western industrial
countries plus Australia and Japan had been greatly reduced since it

first appeared in the 1890s, and how it was still 'a third higher than the average' in the mid-1970s, Also, since 1945 the industrial economies had grown faster than ever before partly due to the elimination of international trade barriers and surpluses of underemployed agricultural labour. Technical innovations had generally seemed to follow rather than to lead growth. Some other countries had overtaken the first ones to industrialize in terms of ability to innovate. Even after recession began biting seriously in the mid-1970s rates of productivity growth were still very high by pre-1960s standards. The first energy crisis was only one reason for the recession. Indirectly it had made Western governments over-cautious about the idea of concerted and controlled action to stimulate demand so as to reduce unemployment.

Economists do not agree upon the reasons and solutions for the relative decline of British manufacturing in particular and of Britain's economy in general. What is clear is that many of the recent benefits of North Sea oil have been used to finance unemployment and that there has been a vicious circle of poor competitiveness, high prices and low sales, low rates of investment, low productivity growth, high labour costs, lack of innovation and relatively unsophisticated products.

By 1983 the value of Britain's imports of manufactured goods exceeded that of her exports of such goods for the first time in over a century. This has reduced the country's ability to pay for foreign food, raw materials and fuel. The country continues to sell more services than it buys from abroad, but its share of the world market for services (which include the skills of some engineers as well as financial services) has also declined. Britain has long been a major centre for such financial services as insurance, banking, brokerage and commodity trading, albeit slowly declining compared with others. Yet although employment in manufacturing has been in decline for most of this century, with that in services tending to rise (for much longer), the economic weight and significance of services has often been exaggerated by apologists for Britain. Productivity gains in services, especially those which are not closely related to manufacturing, are usually far smaller and slower than in manufacturing. British manufactured exports have usually been worth twice as much as the country's exported services. Although the value of the latter has long exceeded that of imported services, the resultant surplus offers only a very limited ability to finance a deficit in manufactures. Many service jobs rely on manufacturing. They support or are supported by it; many jobs in education, distribution and health are in these categories. Indeed the majority of engineers (even) work outside manufacturing (Engineering Council, 1983), and the complexity of the issue of what is a truly productive occupation can easily be illustrated when we think of medicine, electricity supply, education, law enforcement, accountancy, refuse disposal or politics. What we do know is that high standards of technical competence and of living generally go together, and that the wealthier the society, the greater the tendency for machines to replace human labour and vice versa.

World Economic Development and the Implications for Britain

The growing interdependence or internationalization of the world's economy since World War Two can be seen in Table 4.4:

Table 4.4   *The World: Population, Production and Exports by Region, 1950–70 (1950 = 100)*

Region	Population	Gross National Product	Industrial Production	Exports
World total	146	270	280	385
North America	137	210	250	295
Europe	118	260	310	470
Soviet Union	135	435	700	740
Latin America	175	250	300	195
Asia	152	325	820	440
Africa	159	n.a.	n.a.	305

Source: Pinto and Knakal (1973).

Although these data suggest that Europe, the Soviet Union and Asia were the main success stories up to 1970 events were really more complex. When all the figures are adjusted for population growth, production per head in the developing countries can be seen to have grown much more slowly than the uncorrected figures suggest. In fact the developed countries actually built on their 1950 lead in production and in trade in volume terms. The developed 'Western' countries of Europe and North America's share of world trade grew from 61 per cent in 1950 to 65 per cent in 1976. The socialist countries' share rose from 8 to 10 per cent. The developing countries' share fell from 31 to 24 per cent, and except for oil production it would have fallen a lot more. In fact some left-wing writers have suggested that world trade increasingly bypassed the underdeveloped countries after

1945 (cf. Sideri, 1972; Frank, 1980). Thus it is argued that their trade with the developed countries and between themselves both declined quite sharply: the developed countries increasingly traded amongst themselves. However although there is some truth in this argument it tends to assume, rather implausibly, that the relevant events were the product of planning and concerted action by the developed countries. The argument also underplays the fact that the developed countries are simply much more technically sophisticated than the more under-developed ones, sometimes increasingly so, meaning that the benefits of contact and trade between them can be reduced. Finally, the argument often neglects the fact that a fair number of once developing countries are now developed, including several European and Middle Eastern ones after 1945 and more recently several around the Pacific Basin. The latter, the so-called Pacific Rim countries, include Japan and, in many accounts, the West Coast of the United States, as well as such 'new industrial countries' as Hong Kong and Indonesia (Smith *et al.*, 1985). Their often rapid growth, along with that of industry in China and India, is of great importance, implying as it does a major shift of the world's centre of economic gravity from the countries around or near the Atlantic to those around or near the Pacific.

The socialist economies of Eastern Europe have grown very signi-ficantly since 1945 (though more slowly since the mid-1970s) in terms of output and foreign trade, albeit from a lower base than the West. Their increasing involvement in world trade was connected with the pursuit of *détente* between East and West, and this has often led Eastern bloc countries into politically inconsistent postures and actions. Since the late 1970s manufacturing output from the Communist countries, taken almost as a whole including China, Mongolia, North Korea, Vietnam and Albania as well as those of Eastern Europe, has not been growing much faster than manufacturing output from the market economies (both industrial and less industrial) (NIER, 1985). The impact of communist China's economy on the world has generally been small because the country has tried to be self-sufficient. Even so policies have oscillated between modernization through peace, investment and foreign trade on the one hand, and socialist purity implying ideological conflict, support for foreign revolutionary movements, military spending and relative economic isolation on the other. In recent years, especially since President Nixon's visit to Peking and the 'ping-pong diplomacy' of 1972 China's main trading partners have been Hong Kong, Japan, the USA and West Germany. Policy changes notwithstanding, China's foreign trade has grown massively since the 1960s and most of it has been with capitalist countries. China's exports of manufactured goods have long been worth more than those of food, or of oil and other raw materials, and their importance continues to grow.

Since the 1960s most of the industrial countries have been experiencing a relatively depressed phase, partly similar to ones experienced between the mid-1870s and the mid-1890s, and the mid-1910s and the mid-1940s. The expansionary quarter-century after 1945 witnessed some neglect of primary extractive sectors and, in some countries such as the USA and Britain, tendencies to take for granted or neglect technical education and training. Initially, some have argued, there was overinvestment producing over-capacity in manufacturing; later there was overspending on public services. Some of the developed countries may have grown too quickly for their own good, and sometimes too selfishly in relation to some of the developing ones. The latter often became unable to buy from them, so that both major parts of the world's economy tended to stagnate. Recent investment in the developed countries has often sought to cut labour and costs, not to expand capacity.

Politically the developed Western countries have tended to move to the right in a defensive reaction to recession, with left-wing parties or important parts of them moving to more extreme positions. Politicians have failed to persuade electorates to moderate inflationary pay demands, and are now so afraid of inflation that they are unwilling to reflate their economies to reduce unemployment. Advisers to governments, confused by the growing and unpredictable interdependence of national economies, only felt able to advocate 'prudence'. Welfare cuts and increased military spending have resulted, along with the exporting of national balance of payments deficits to the poorer countries, some of whose debts have reached severe crisis proportions. Some of the costs and problems of employing people in the developed countries were also exported to the poorer ones, who offered tax reliefs, direct state subsidies, and cheap and often politically repressed labour to incoming foreign employers.

Defensively irrational fears of the spread of democratic practices outside the Soviet bloc and also more understandable fears of the spread of communism due to the latter's influence seem to be connected with these changes, although the economic deeds – as opposed to the rhetoric – of the Comecon countries seem designed to fit in with the West's aims, not to subvert them. Also the Soviet military expansion since the mid-1960s is probably best interpreted as a catching-up exercise, which NATO could always surpass because its members have about one-and-a-half times the population and more than twice the productive capacity of the Warsaw Pact nations. It is also hard to justify all the fears about shortages of raw materials which have worried the West. A wide range of views has been expressed on the availability of raw materials and energy, but at least one expert has expressed the view, a little optimistically perhaps but after extensive consideration of evidence, that given some recycling and reasonable caution towards expansion, 'non-fuel minerals will . . . be available for the indefinite future', and that even fuels will only become a serious problem if we lose our so-far well proven ability to develop substitute and new resources (Tanzer, 1980).

Some long-term interregional economic trends may not favour Britain and the rest of Europe. The recent economic strength of Europe, North America, Australasia and Japan seems to have originated in the opening up of the New World from the fifteenth century onwards. This involved intense naval and military competition between the European countries, later drawing in the USA and Japan, all of which overtook older civilizations in Africa and the Asian land mass in technical and military terms. Today sea power and communications are not quite as important in relative terms as they were two centuries and more ago because of the development of communications on land and in the air. Surface naval power is nowadays vulnerable to attack from land-based aircraft and missiles. Sea transport remains cheap and efficient compared with other forms but air and land transport have sometimes replaced or complemented it. This suggests that the maritime West, especially Western Europe with a very lengthy coastline made up of peninsulas (or of islands like the British ones), may be in decline relative to other more continental parts of the world (Guha, 1981). However while British power has declined greatly (in relative terms) since the nineteenth century and while the USSR has grown faster since 1945 than the USA or most of Western Europe, the case of Japan and of other Pacific communities hardly fits Guha's argument very well. Nor does that of the USA which is a maritime power in many important ways, albeit a continental one in others. In fact the importance of sea communications, power and transport has been and still is increasing in absolute terms. Moreover the values of Western countries tend to be activist, instrumental and growth-oriented, whereas more passive philosophies still tend to prevail in India, China and many parts of Africa and the Middle East.

In these circumstances however, are Britain's economic difficulties that serious? Offshore oil and gas may act as major crutches, perhaps for another generation. It is also possible that Britain will become a net exporter of foodstuffs in the 1990s, whereas about 50 per cent of Britain's food had to be imported in the 1950s and 1960s. But manufactured goods are still being imported at an ever-growing rate.

Manufactures and semi-manufactures accounted for 4 and 14 per cent of British imports in 1950: the equivalent figures for 1979 were 36 and 29 per cent. The earlier figures were unusually low in the aftermath of the Second World War, but a jump in import penetration from nearly 10 per cent in 1963 to 28 per cent in 1982 was exceptionally worrying, especially when we note that a surplus of exports over imports of 100 per cent evaporated in the same period.

Many other kinds of statistic – on declining rates of profit, rising burdens of taxation, on growth in the proportions of the labour force working in 'unproductive' government, declining investment in important sectors, losses of manufacturing sectors (in whole or part), and slow or abnormal productivity growth – could be cited to paint a very pessimistic if not a hopeless picture. It would not be altogether unreasonable to describe the situation as a war in which manufacturing and construction, in the front line, bear most of the hardest burdens, while those in the rear, governments, consumers and voters, look after themselves.

Informed commentators appear to agree broadly that at least three major policies are needed for the tide to be turned for Britain. They should be: first, more vigorous and consistently helpful behaviour by governments towards manufacturing and construction (in investment, regional policy and taxation, for example); second, sweeping *and* careful educational changes designed to produce practical makers and doers who are broad in their abilities and sympathies, for all levels of employment (*not* factory fodder in any shape or form); and third, a highly participative climate in all sectors of employment (perhaps abolishing much of what we know as 'management' and 'industrial relations').

According to at least one writer failure to achieve these things will mean that, for most Britons, 'post-industrial society will be a replay of the Hundred Years War' in which 'they will [be] on the losing side', mainly as participants in a 'vast backwater economy . . . where unemployment, menial work, crafts, moonlighting, barter and brigandry are the standard forms of everyday life' (Bellini, 1981).

5 The Legal and Political Context of Engineering

Introduction

This chapter concludes the first part of the book and therefore brings to a close our arguments which highlight the problems of engineering. In the first section of the chapter we examine some of the legal constraints affecting employment in engineering in Britain. These are seen to have been influenced by broadly defined political forces. We next examine the varied roles which engineers can and do play in political life in different countries. Then we turn to the relationships between state power and engineering and in particular to the very different ways in which industrialization has occurred in different settings under the influence of different political arrangements. Finally we consider an issue which may increasingly concern engineers, politicians and social scientists in the affluent industrial countries. Do the high living standards which successful engineering produces *necessarily* undermine the standing and power of the engineer? In other words is the problem that we have defined as engineering's lack of centrality in the British social system a consequence of Britain's engineers' earlier successes?

Engineering, Employment and the Law

British engineers need to know a lot more about employment and other kinds of legislation than their counterparts did twenty years ago. There were five general Acts of Parliament regulating employment passed from 1950 to 1959, and sixteen from 1960 to 1969. There were thirty from 1970 to 1979 and a further eight from 1980 to 1982 (Hepple, 1983). There was also considerable growth in the number of more specialized statutes, and in case law concerned with new rights enforced through the industrial tribunals originally set up in 1964, in connection with training. Governments have shown increasing concern with training, with the organization of collective bargaining, and with inflation, often using the law in the cause of change.

A major area of expansion has been in the legal employment rights of individuals. Most rights are enforceable, at least in theory, in the national network of industrial tribunals which now deal with many issues. One of the most important is the right not to be unfairly dismissed. Most types of employee may only be dismissed fairly if they are unable or unqualified to do a job, if their conduct is unsatisfactory, if their jobs are genuinely redundant, or if there is some other valid legal or other reason which justifies dismissal. Someone who is pregnant or an active trade unionist, for example, cannot be fairly dismissed for that reason. If an industrial tribunal decides that an individual has been unfairly dismissed it can order reinstatement or re-engagement, or require a limited amount of financial compensation to be paid.

New employment laws and the codes of practice associated with them have given a substantial stimulus to the introduction of formal disciplinary and grievance procedures by employers, and to the development of legal expertise in personnel management. The idea of an individual employment contract between theoretically equal parties has become central to the understanding of British industrial relations. This idea came late to Britain; in fact many employment laws were in existence before its widespread acceptance and current employment law is both complicated and patchy in coverage. Some laws are enforceable through the criminal courts, others through the civil ones, and others by administrative bodies. However, and as we argue at more length a little later on, labour law in the United Kingdom has never, except in wartime and immediate post-war periods, played any *regulatory* role in industrial relations.

Some employee rights are statutory; others form part of individual contracts. Although formally defunct, the traditional definition of employers as masters and employees as servants effectively persists as a kind of legal precedent which, along with other things, tends to make judges favour employers over employees. The relatively new idea of free individual contract ignores the fact that the main terms and conditions of employment of most employees were originally and still are to a large degree determined by collective bargaining. In practice most terms and conditions are determined by a mixture of collective and individual arrangements. Individuals are not legally bound by agreements made by their unions; also such agreements generally apply to union members and non-members alike.

A contract is a bargain to which there are two sides. Each side or party does or promises to do something in return for action or a promise of action from the other. In a contract of employment, an employee agrees to work for an employer and the employer agrees to pay a salary or wage in return. Once someone agrees to work for a certain sum, a contract of employment exists. The contract does not have to be in writing or formal in some other way.

Many employment contracts are not written ones. Each contract of employment is however underpinned by various factors. Thus employers' rules and regulations such as disciplinary rules and various other items in company handbooks are deemed to form part of the 'implied terms' of the contract. Collective agreements negotiated between employers and trade unions form the 'express terms' of the contract: they detail current terms and conditions of employment. However custom and practice and unwritten rules of a workplace or economic sector often form part of the implied terms and they can even override the 'express terms' of particular collective agreements. The common law, judge-made case law, nowadays fills in gaps left by statute law when judges interpret existing statutes in relation to particular cases. Common law places general obligations on both sides which are 'implied' in every employment contract, often irrespective of whether or not something has been written down or stated explicitly.

The common law duties of each side consist of assumptions or rules based on what judges deem their mutual obligations to be. Typical common law duties of employers are to pay agreed wages in return for willingness to work, to obey the law including that concerning health and safety with regard to employees, to treat employees with proper courtesy and, in some cases, actually to provide work. Typical common law duties of employees include being ready and willing to work and to co-operate with employers in the performance of their legal duties, to take reasonable care in the exercise of services to employers, to obey all reasonable and lawful orders concerning times, places, nature and methods of working, not to disclose confidential information or to place themselves in positions in which their interests may conflict with those of employers. The statute law affects contracts of employment when it determines general rules about terms and conditions of employment on such things as hours of work and pay, and health and safety.

Contracts can be ended in various ways. One is by 'due notice', given under the terms of contract by either party, and subject to the employee's right not to be unfairly dismissed. Another is by 'frustration', whenever something occurs to prevent the continuous performance of the contract or radically to alter its basis. Obviously death is one such instance; so too may be imprisonment or lengthy illness. There is also 'repudiation', when one of the parties breaks an important term in the contract. Examples of this on the part of the employer may include victimization or demotion of an employee, which can entitle him or her to claim 'constructive dismissal'. Similarly if an employee breaks his contract through an act of gross misconduct, for example, the employer may dismiss him without notice under common law. Contracts cannot be changed by only one party to the bargain, i.e. unilaterally. This can be very important if managements want employees to do new kinds of work or to transfer them. If the existing contract does not allow for the necessary flexibility, either expressly or implied, unilateral action could mean a repudiation or other breach of contract.

Like all types of law in Britain, employment law is produced and regulated by various legal agencies, terms and institutions. The three main sources of law are Parliament, which creates statute law; judges, who create common law to fill the relevant gaps in statute law (although much common law existed before statute law was extensively developed); and the European Economic Community (EEC). The EEC can make regulations which are legally binding on member states even though laws may not have been passed by individual national governments, and it can issue directives which are put into practice by national governments passing appropriate laws applicable to their own peoples. EEC laws override national laws when there are inconsistencies between both.

Of the two types of law in Britain, criminal law has been relatively unimportant with regard to employment matters although 1980s employment legislation has changed the position somewhat, affecting such matters as picketing and some other types of industrial action, and health and safety. It is enforced by the police, and for more serious crimes by the Director of Public Prosecutions in England and Wales and the Procurators Fiscal in Scotland, along with the courts, and penalties consist of fines and/or imprisonment. Civil law, which relates to disputes between individuals, usually involves some breach, by either party, of its contractual obligations to the other. A party bringing a civil action must establish that a wrong has been suffered for which damages are payable. In legal terminology a civil offence for breach of contract is called a 'tort' or 'delict' in Scotland. Most cases involving employment law are civil ones of this type, being concerned with disputes between employers and employees about aspects of their mutual contractual obligations.

The system of the courts which enforce both criminal and civil law and the details of types of law vary between England and Wales on one hand and Scotland on the other. The legal system is served by three main types of legal professional. Solicitors give legal advice and represent their clients in the lower (e.g. county and magistrates' or, in Scotland, sheriff's) courts and in industrial tribunals. In the higher courts however barristers take over the job of representing clients from solicitors, from whom they receive instructions. Barristers cannot deal directly with the public. Judges apply statute and common law and develop case law; they are drawn from the ranks of barristers.

Industrial tribunals, part of the civil law system, form the most important part of the legal structure concerned with employment law. Typical concerns of industrial tribunals include unfair dismissal, redundancy payments, time off for trade union activities, and racial and sexual discrimination. They sit throughout the UK and they are often described as informal courts of a kind. Each consists of a panel with a legally qualified chairperson, and two lay members, one nominated by the Confederation of British Industries (CBI) and the other by the Trades Union Congress (TUC). The parties do not need to be legally represented, although it helps if they are or if someone else represents them, such as an expert volunteer from their local Citizens' Advice Bureau (CAB). If either party feels that an industrial tribunal decision is wrong because the law has been incorrectly applied or interpreted, it may seek a remedy from the Employment Appeals Tribunal. Appeals from it go through the more traditional channel of the appeals courts (Court of Appeal, Civil Division, in England and Wales, Court of Session, Inner House, in Scotland).

We have already noted how, in practice, British labour law has normally not played any regulatory role in industrial relations. Such legislation as existed up to the post-1945 period could best be regarded as supportive and/or enabling legislation, i.e. legislation that provided legal and institutional support for what was essentially a *laissez-faire* system of collective bargaining.

However it was not that the legislation was unimportant. The Conciliation Act (1896) and the Industrial Court Act (1919) provided the basis for a whole range of state agencies and services in the fields of conciliation, arbitration, mediation and inquiry, and these services were valued by the practitioners of their day. Yet the essence of these services were that they were largely voluntary, except of course for Committees and Courts of Inquiry where a certain minimum amount of compulsion was available if necessary, and whether or not people used the services or not was entirely up to the parties themselves.

The Wages Councils legislation was of a different order. From 1909 the state accepted on purely social grounds the necessity of intervention in the basic terms and conditions of employment for approximately 4 million workers in situations where they could not sustain effective organization for bargaining purposes and, in the majority of cases, neither could the employer. So a form of state-sponsored, state-supported collective bargaining came into being as a temporary measure. At least that was the original idea. In fact the need for such provision was accepted as being essential until quite recently.

The employment legislation of the 1980s arose for quite different reasons. It is partly designed to weaken the trade unions, whose legal rights and powers have always been suspect. The method is not an open attack by the government on the unions in which the government passes legislation which creates crimes for which people can be charged, tried and, if found guilty, punished. Instead, what the legislation does is to create torts (legal wrongs which are not crimes) and then the action, if any, to be taken is left up to any of the parties who may wish to take it, employers, trade unions and their members being the main parties. In other words, if no one feels the need or the urge to take action, then the law as such does not exist.

A second feature of the legislation has been a relative one-sidedness. To encourage the parties to act, the device of the injunction is freely available to the 'injured' party. This often means that, in practice, an employer does not have to mount a full legal case against an 'offending' union. All that the employer has had to do is to apply to the courts for an injunction (an interdict in Scotland) restraining a union from doing X (whatever X might be) and the courts will – it seems – usually grant such an injunction. In this way, the unions have often lost the day whatever the legal or other rights or wrongs of the situation may be. If the union fails to obey the injunction the employer can notify the courts and the union will be called before the courts to answer a charge of contempt. If the union does appear, then it is (in effect) back to square one. If it doesn't it will be held to be in contempt of court and fined, often very heavily. If it pays the fine, it still has to obey the injunction. If it doesn't, then the courts may proceed to seize some of the assets of the union. Since the legislation of the 1980s was passed the device of the injunction has been most freely used by employers, but it is nevertheless just as readily available to the unions. Because of past experiences of the legal system and their natural tendency to be somewhat anti-establishment the unions have been reluctant to use or co-operate with the courts. Even so unions and union members have recently used the injunction device in a few cases.

The legislation which lays down rules of employment, restricting managerial power, is often called regulatory legislation. Examples of items covered include wages, hours, safety, training, dismissal and redundancy. It is mainly used to determine things which usually lie within the scope of collective bargaining, such as wages and hours, and to protect unorganized workers. Otherwise it regulates such things as safety and training, which are not typically dealt with in collective bargaining. Auxiliary legislation (on the other hand) is concerned with the ways in which agreements are made, rather than with their content. It is used to promote negotiation, agreements and the observance of agreements.

In the nineteenth century, regulatory legislation was mainly concerned with hours of work, safety, and the calculation and methods of paying wages. Early in this century it secured minimum wages for unorganized workers by setting up the Wages Councils. More recently there has been legislation on equal pay (1970), sexual (1975) and racial (1976) discrimination, and on job and income security (various Acts since 1963). Conservative governments since 1979 have removed some of the aspects of individual rights in employment, although most remain. Some long-term trends may favour a further expansion of individual employment rights. For example multinationals press for unification of employment practices in their host countries in the cause of equality of competition; the EEC with its often higher-than-British employment standards, and British pressure groups, may also press for changes on such matters as age discrimination, job sharing, study leave and hours of work.

Collective labour law includes auxiliary legislation, determining the legal framework for collective industrial relations. In practice however it overlaps considerably with the law governing individual employer–employee relationships. It covers collective bargaining relationships, inter- and intra-union relationships, and the forms that conflict can take, such as strikes, picketing and working to rule. Collective labour law has generally played a largely supportive second fiddle to a voluntary system of collective bargaining between employers and unions whereby agreements are informal.

In the 1960s, as at various other times, poor industrial relations were a major scapegoat for relative economic decline. Numerous attempts at reform and restriction followed. The important Donovan Commission (1965–8) wanted more orderly and efficient employment relationships. It sought voluntary reforms of collective bargaining, to be underpinned by a legal framework concerned, among other things, with union recognition and unfair dismissal. Labour government proposals of the late 1960s and the Conservative Industrial Relations Act of 1971 embodied similar reformist aims, strengthened with legal teeth in some cases. They also embodied more restrictive aims, opposing unofficial strikes and the closed shop for example. Some elements of the 1971 Act were very restrictive, seeming to undermine the basic collective nature of trade union organization, and to threaten the freedoms of the labour market. It was formally repealed in 1974 after Labour returned to office, following union opposition and apparent management indifference.

Between 1974 and 1979 Labour governments initially offered the unions a return to completely free collective bargaining, and action on prices, incomes, rents and industrial democracy in exchange for pay restraint in order to combat inflation. This became known as Social Contract. Unions and employers would draw up 'planning agreements' concerning the future of businesses, investment and marketing policies, and so on, and the National Enterprise Board was set up to increase public ownership, to encourage investment in industry, all as part of an attempt to produce a more efficient, planned, egalitarian society. However policy became increasingly focused on pay restraint, eventually helping to lead to the so-called 'winter of discontent' of 1978–9, against a pay policy applied very rigidly to lower-paid public sector employees (Kelly, 1983).

After 1979 Mrs Thatcher's Conservative governments sought to enhance individual freedoms at the expense of collectivism, and to curb apparent excesses of union power. The closed shop and picketing were two of their more important concerns. Recession and unemployment made their tasks easier. Thus the early 1980s witnessed a major reduction in the number and duration of strikes and of other forms of collective action. Important exceptions to this however include a number of spectacularly unsuccessful disputes in the public sector, which have generally illustrated the declining power of the trade union movement under Conservative governments in a period of recession.

The Conservatives' most important measures were the Employment Acts of 1980 and 1982 and the Trade Union Act of 1984. They were probably not as radical or comprehensive as the 1971 Industrial Relations Act, but they were nevertheless described as attempts to undermine employee power and rights and the principle of collective bargaining. The 1982 Act negated employment contracts which stated that only union members could do certain tasks, or that employers had to recognize and negotiate with particular unions. Together the Acts of 1980 and 1982 effectively nullified the law's longstanding support for the right to belong to a union (which had previously been beginning to be seen as an implied duty), and for the development of collective bargaining. The right not to belong to a union was strengthened. It was now much harder to dismiss non-trade unionists even when a closed shop agreement operated. Closed shop agreements were no longer allowed to operate retrospectively to coerce existing non-members to join unions, although they rarely did so in the past in any case. The trend was to support individuals against unions and, in principle at least, to support an erosion of trade union organization. The arguments supporting the legislation emphasized the case for democracy within unions and a desire to 'give them back to their members', and to reduce the power of permanent officials. Before 1980 such legislation as existed affecting the closed shop was, in effect, broadly permissive. Safeguards for non-members were widely known to exist within voluntary agreements between employers and unions. In spite of some politicians' statements to the contrary there always has been a right not to belong to a union, both before and after 1980.

The right to strike was also effectively curtailed. Since 1982 it has been harder for an employee dismissed while on strike to claim unfair dismissal on the (only relevant) ground of victimization. It has been more difficult for strikers' families to receive Supplementary Benefit. The 1980 and 1982 Acts cast doubt on the idea that those who withdraw their labour have not wronged their employers. The right to picket was severely curtailed (in theory): unions were advised that the number to be present at an entrance to a workplace should not exceed six, some kinds of secondary picketing were forbidden, and picket organizers were made directly responsible to the police. Disputes which took place for reasons which were not 'trade' ones, such as political strikes or ones arising out of disputes between unions and/or workers about the closed shop, demarcation, recognition and recruitment, were made either illegal or barely legal. In the case of disputes which were not limited to relationships between particular groups of employees and their (and only their) employers, the trade unions lost their right to immunity from prosecution by employees under the civil law. This immunity has also ceased to exist in the case of disputes called without a secret ballot of union members. It was made theoretically possible for the first time since 1974 for unions to be made responsible for their members' or officials' actions. The main details of the Conservative Acts of 1980, 1982 and 1984 follow.

The 1980 Employment Act was mainly concerned with trade unions, industrial disputes, and employees' individual rights. The Act introduced changes to the Employment Protection Act of 1975, the Trade Union and Labour Relations Act of 1974, and the Employment Protection (Consolidation) Act of 1978. With regard to the trade unions, the main provisions were as follows. Employers with more than twenty employees were made to provide facilities for secret trade union ballots on their premises (this referred only to unions recognized by employers). Funds for trade union ballots could henceforth be provided by the Government. The Advisory, Conciliation and Arbitration Service (ACAS) was no longer to be involved in recognition disputes between trade unions and employers. Closed shops were required to have 80 per cent of those covered by voting for them (or 85 per cent of those voting). A dismissal was to be deemed unfair if an employee was sacked for not belonging to a trade union on conscience grounds. Employees unreasonably excluded or expelled from a trade union in a closed shop situation were to be paid compensation from the relevant trade union if an industrial tribunal so decided.

With regard to industrial disputes, picketing was only to be lawful if it occurred at the employees' workplace. Secondary picketing was thus restricted (unless a supplier or customer of an employer was the subject of the picketing). Picketing outside this restricted area could lead to the trade union losing its immunity from legal action for damages by the aggrieved party. A new code of practice on picketing was introduced. With regard to individual rights, there were some minor amendments concerning guaranteed payment of wages. On unfair dismissal, the service period required in the case of employees of small (under twenty employees) organizations was raised from one to two years, lengthening the period during which a 'new' employee could be dismissed more or less at will. There were also changes in the regulations relating to the ways in which industrial tribunals carry out their functions concerning unfair dismissals. On maternity rights (see also below) employers became entitled to more firm information than was previously needed about whether a pregnant employee/mother would return to work. The 1980 Act also weakened the principle of 'fair wages'. More generally its major effects were to reduce the legal immunities of unions, and the protection given by the law to employees of small firms.

The 1982 Employment Act developed the 1980 legislation in three main areas: the right to strike, trade union immunities and closed shops. On the right to strike, there were changes in the definition of a trade dispute, and the boundaries of lawful industrial action were redrawn. Henceforth industrial action had to be concerned with terms and conditions of employment, physical working conditions, sacking, supervision, disruption and trade union rights. Disputes concerned with the following were no longer lawful: demarcation, disputes overseas where these did not affect the working conditions of British employees and disputes between employees and an employer where the latter has no dispute with his/her employees (sympathetic strikes, etc.); and employees involved in occupations and sit-ins were no longer immune from legal action. On trade union immunities, trade unions were now liable to be prosecuted for unlawful industrial action when it had been authorized by their officials (this applied to *official* disputes, and many unions were thus likely merely to declare disputes unofficial and thus to escape liability). Damages, however, were limited to £10,000 for unions with less than 5,000 members, rising to a maximum of £250,000 for unions with over 100,000 members. On closed shops, the position of non-union members was strengthened. Employers were now subject to very real financial penalties for dismissing individuals who refuse to join unions. Contracts which stated that only union labour may be used were now to be void. So too was any term in a contract which said that the parties must recognize or negotiate with a trade union. Employee involvement statements were required from directors of large companies, who had to include a statement in their annual reports about what had been done to introduce or extend employee involvement in company affairs.

The 1984 Trade Union Act required secret ballots to be conducted on a range of trade union matters. Since October 1985 trade union executive members have had to be elected by secret ballot, at least once every five years. Secret ballots are now required before strikes, if the relevant trade union's immunity from legal action by the employer is to be preserved. Also ballots have had to be held before March 1986 for those unions with political funds to see if their members want the political funding to continue.

Labour's Employment Protection (Consolidation) Act of 1978 consolidated most individual employment rights. It was this piece of legislation which the Conservatives concluded as they sought to reduce those rights, with the professed aim of making it easier and more attractive for employers to take people on. The Conservatives' 1971 Act was much more supportive of collective industrial relations than their 1980 and 1982 Acts. The last two were introduced when mass unemployment had weakened unions and when employer organization increasingly had a multinational, multi-plant character. To follow the 1984 Act the Government was considering legislation to encourage mergers between trade unions, to abolish Wages Councils and to curtail the right to strike in parts of the public sector.

During the first half of the 1980s the basic political and philosophical conflicts have been between Conservative adherents of sound money, individual freedom, self-help (and devil takes the hindmost) on the one hand, and on the other the collectivist, corporatist philosophy of the Left, which wants a strong, disciplined union movement to be a keystone of a democratically planned economy. The SDP–Liberal Alliance ostensibly seeks a middle way which, perhaps even more than the Conservatives and Labour, can be presented as being simultaneously effective, radical and caring. However its sympathies in industrial relations matters are seen by many trade unionists as tending towards the Conservative position. Although employment law has been something of a political football – not least because it is often a convenient diversion from more basic issues – many of the legal provisions established since the 1950s remain intact, and many of the partly-eroded ones are still effective to some degree. A basic outline of some of the main current legal provisions affecting employment follows.

After engaging someone for the first time an employer has up to 13 weeks to provide them with a *written statement*, containing at least the main terms and conditions of employment, and an outline of grievance and disciplinary procedures. This statement usually forms an important part of the contract of employment. If the statement contains all details of terms and conditions of employment, 'express' and 'implied', including all relevant rules and duties, it may virtually comprise the whole contract of employment. Dismissal due to redundancy often means that a compensatory lump sum is payable. Sometimes employees laid off or on short time for lengthy periods can also claim. The amounts payable depend on age, length of service, and rates of pay. Unions have to be consulted about proposed redundancies and employees working under notice of redundancy have rights to time off for job hunting or for arranging training.

Most female employees have certain maternity rights: they have to be given reasonable time off (paid) for antenatal care, they cannot be fairly dismissed because of pregnancy, and there are rights to receive some maternity pay and to return to work at any time for up to 29 weeks (sometimes more) after giving birth. Rights exist both to belong (and to be active in) and not to belong to a trade union without being dismissed. Most employees have to be given itemized pay statements whenever they are paid, and most are theoretically eligible for guarantee payments when complete working days are lost because there is no work to do or for any other reason. Many employment rights, including the one not to be unfairly dismissed, depend on a qualifying period of continuous employment. Its length depends on the right in question as well as other factors such as strikes, size of employing organization in terms of the numbers employed, maternity leave, and those apparent and real changes of employer which occur when establishments change hands but jobs remain more or less the same. There is also a right for time off for public duties such as those of members of juries, governing bodies of educational institutions, local and other public authorities and statutory tribunals.

An unfair dismissal, as noted earlier, can occur when what the law regards as a substantial reason is not applicable. Grounds which justify dismissal are lack of ability or qualifications to do a job, misconduct, redundancy, legal duties or restrictions preventing employment being continued, or another substantial reason. Rights concerning termination of employment also cover notice, pregnancy, trade union membership, criteria for selection for redundancy, sick and holiday pay, fixed-term contracts, and information concerning reasons for dismissal.

The Health and Safety at Work Act of 1974 produced a broad framework for the development of a comprehensive, integrated system of law and practice on health and safety at work. Points covered by it include employers' safety policies, safety committees, first aid and fire-fighting facilities, fire drills, safety representatives, records and procedures. Employers are also variously encouraged to provide training and, as noted earlier, to refrain from discrimination on grounds of race or sex (but not age). They may only pay employees by cash or cheque, and they have to give recognized trade unions information relevant to collective bargaining.

Details of all of these and other features of employment-related law can be obtained from such personnel management texts as those by Armstrong (1984a), Torrington and Chapman (1979), Thomason (1981), and especially from specialist studies concerned with employment law such as Field (1982), Lewis (1983), Whincup (1983) and Selwyn (1982). Several chapters of the industrial relations reader edited by Bain (1983) are also highly useful as guides to the background to legislation. Commercial, company and contract law also affect the engineer's work; so too do laws affecting sales and insurance. Texts on managerial aspects of engineering such as those by Turner and Williams (1983), and on company law (Charlesworth and Morse, 1983), the sale of goods (Marshall, 1983; Dobson, 1984), commercial law (also Marshall, 1983, and Charlesworth, Schmitthof and Sarre, 1984), delict (Marshall, 1982), tort (Baker, 1976, and Salmond, Heuston and Chambers, 1981), and consumer protection (Harvey, 1982) may also usefully be consulted by engineers.

Engineers in Political Life in Britain and Abroad

Public figures and social commentators have often compared British engineers with such people as lawyers, academics and journalists in

Britain, and with other engineers abroad, to criticize them for an apparent lack of involvement and interest in politics and government. Jimmy Carter, albeit unusually for an American President, was an engineer, whereas no British Prime Minister has been one. Several major world leaders such as Kosygin in the USSR and Giscard D'Estaing in France have been engineers. Hinton (1970) pointed out that the number of British engineer MPs did not reach double figures and how the Administrative Class of the Civil Service had hardly ever recruited engineering graduates. Snow (1966) argued that British engineers were politically more passive and conservative than arts or natural science graduates. Studies by Gerstl and Hutton (1966) and Halsey and Trow (1971) supported this point insofar as samples of mechanical and academic engineers were slightly to the right in the political spectrum.

Lack of political interest and involvement by British engineers is sometimes associated with an assumed lack of involvement in non-technical aspects of management. Two Americans have argued quite impressively around the idea that the engineer's role in society might be a more strategic one. Veblen (1921) as we pointed out above felt that the engineer's usually close contact with life's realities would eventually, as engineers became more powerful, make political decision-making in industrial countries more sensible and honest. Merton (1957) was more pessimistic about what he called the (American) engineer's 'trained incapacity', which seemed to take the form of a narrow problem-solving approach to complicated human affairs. Generally sociologists have produced a good deal of high level speculation about the impact of technical change on political and economic decisions. With other social scientists, they have written much about the administration of government support for innovation and on the role of natural scientists – but rarely engineers – in government and politics.

Debates about public support for science and technical innovation have focused since the 1970s on the increasingly 'applied' or 'customer-oriented' character of publicly-financed research and development (Rothschild, 1971). The most basic issue to concern writers is the degree to which scientific research should be justified on economic and social grounds. Price (1965) argued that if a historical perspective is adopted it can be seen that most engineering and scientific achievements have occurred independently of each other. Each was a cumulative system which relied heavily and on the whole exclusively on its own past achievements. Links existed but were usually tenuous. Clearly industrial growth can take place whether a country supports basic science or not; the rapid development of Japan since 1945 is often cited in favour of this view. Japan's *more recent* investment in scientific research cannot be used to contradict it.

Many natural scientists have however argued, especially when research funds are being sought, although with some justification, that engineering does depend on fundamental science. It has been argued for example, and reasonably so, that many post-1945 technical developments derive at least partly from fundamental research done in the 1930s, in quantum mechanics, optics, particle physics and electromagnetic theory.

It is of course true that engineering students study a great deal of natural science, and that technical development is often clearly reliant on the availability of scientific knowledge. Both of these points seem to support the view that engineering needs science. However no one would call a cook, a sculptor or a detective an 'applied scientist' (or an engineer) just because they need to use scientific knowledge in their work.

In some settings the older distinction between basic and applied science has been replaced by one between Big Science and Little Science (Price, 1963). Big Science means either large-scale mission-oriented work – like the US space programmes – with vertical integration of processes from basic research to technical support of operations, or institutional basic research, in organizations which exist solely to do long-term and often very expensive research. Little Science means academic science, usually quite small-scale, with planning relatively short-term. However engineers and former engineers like Tobias (1968) and Hinton (1970) would argue that mission-oriented Big Science is largely engineering and that institutional basic research often is too, depending on its aims; and that even Little Science involves considerable use of engineering skills. These authors tend like us to prefer the simple distinction (made elsewhere in this book) between engineering and science, according to which they differ fundamentally, because whereas *engineers make things, scientists study them.* They also tend to dislike the idea that the flow of information and ideas between science and engineering is normally one from science to engineering. They suggest that there is often little direct contact but that when there is the flows often proceed in the opposite direction to the one in which they are theoretically supposed to go.

The main practical issues for makers of what English speakers rather misleadingly call science policy are as follows. Does spending on basic research represent a worthwhile long-term economic and cultural investment? How much effort and/or money should go into technical and scientific education and training? How far should governments go in dictating their content and direction? In post-war Britain there have been more and more attempts – sometimes quite half-hearted ones however – to make scientific research more mission-oriented and to put more resources into technical education. In most countries scientists want the freedom (and money) to pursue truth, whereas their governments want useful results. In the English-speaking countries, notably Britain, engineers need to assert the separateness of their aims and tasks from those of science. The issue is to raise engineering's status by emphasizing its *own* identity and importance, rather than by trying to borrow status from science. Whether engineers like it or not, this is a political issue.

An old recurring and relevant idea in sociology is the 'technocracy thesis'. It refers to, and usually attacks, the view of the world which assumes that social and political *ends* are non-problematic and more or less universally agreed upon, and that experts should decide on the *means* for attaining them. For example Ellul (1964) suggested that 'technological society' is one dominated by 'technique', in which formal planning and bureaucratic structures stifle spontaneity, individuality and imagination (see also Marcuse, 1964; Roszak, 1970; Habermas, 1971; Sklair, 1973).

The development of different types of engineering and of ways of organizing engineering tasks are to a large extent functions of political decisions. For example the Ministry of Technology was set up by the Labour government in the mid-1960s, to 'spearhead the drive for technological efficiency'. It was abolished by the Conservatives after their return to power in 1970 almost as impetuously as the way in which it had been established, although most of its tasks continued to be done elsewhere in Whitehall. The various employment opportunities and training schemes initiated since the mid-1970s by the Manpower Services Commission similarly present as superficial and half-thought-through – although more recently as improving – attempts to do something about Britain's historically chaotic and inadequate training provision (for the superiority of foreign efforts, see for example Prais, 1981a; Prais and Wagner, 1983; NEDC/MSC, 1984).

Since the First World War the various research councils for agriculture, medicine, the natural environment, science and engineering, and social science have been developed to administer funding of research. A key feature of the system is 'dual funding'. Places of higher education provide research proposals, research directors, working facilities and equipment, and the research councils pay the salaries and expenses of contract research staff. The Science Research Council became the Science and Engineering Research

Council, to acknowledge the important engineering-oriented research done in its name. Proposals to hive off a separate Engineering Research Council were however felt to be going too far, even in the atmosphere preceding and following the publication of the Finniston report in 1980. Suggestions that the Science Policy Research Unit at the University of Sussex be similarly renamed were also not acted upon, although they were apparently taken quite seriously at first. Some of the same increased public concern for engineering was associated in 1983 with the renaming of the Social Science Research Council (SSRC), which became the Economic and Social Research Council (ESRC): more economics and management, less social anthropology and sociology, and blame for not being as precise as physics were all implied. The odd nature of the logic that if something is relatively hard to measure, its study is not truly scientific, seems to have been lost on the politicians who originally suggested the need for a new name in which the 'Science' in SSRC would be replaced by 'Studies' (cf. Glover and Schröck, 1983).

Although the extent to which engineers fill top posts in business and government in the industrial countries varies a great deal, Galbraith (1967) argued that decision-making in all of them was increasingly a group exercise on the part of a 'technostructure' of expert, 'professional managers' of all types, rather than their nominal leaders at board or equivalent level. He cited the dependence of the latter on experts for technical information as evidence, referring too to the size of larger organizations. The organization had, in effect, subverted its formal leaders. However the many critics of Galbraith's view argue that the high cost of knowledge and expertise is simply a function of our wealth and that the wealthy (including in this instance corporations) are still just as able to hire and fire experts as they ever were. Critics also emphasize the point that *where* one has studied, not only *what*, still influences recruitment and promotion prospects quite strongly, and that 'ownership of wealth continues to play a fundamental part in facilitating access to . . . education (for) entry to elite positions' (Giddens, 1973: 263–4).

The true technocrat is probably best exemplified by the French 'diploma engineer', especially one who is a product of one of the engineering *grandes écoles* which recruit many of France's most able school-leavers, and which offer courses widely felt to be more intellectually demanding than university ones. This produces high-powered, broadly educated and practical engineers, commercially, financially, socially and politically aware and skilled, able to act *consciemment*, literally with perception or discernment. Also he 'may employ a career pattern known as *pantôuflage* which involves moving between industry and various government agencies to train him for such roles as "indicative planning" ' (Hampden-Turner, 1983; Glover, 1978a). German or Swedish counterparts are similar in ability level and range of subjects studied. Sweden's equivalents of Oxford and Cambridge produce high-powered engineers and little else; running factories is a traditional route to the top of Swedish society, not just of Swedish industry. The British counterparts of such people are typically arts or natural science graduates, educated in ways not designed to produce a worldly outlook, but a desire to serve the community while setting an example to others (Barnett, 1972), or to be a seeker after truth.

Are most engineers really uninterested in political matters and/or conservatives with a small 'c'? Some past British and American studies noted the tendency of engineers to come from humbler social backgrounds than other university students, and also purported to show that they preferred working with things rather than with ideas or people, that their political and social attitudes tended to be authoritarian and prejudiced, that they were narrowly trained and narrow in outlook, with modest aspirations and a lack of interest in general ideas, activists who sought order and certainty in a rather single-minded way, who regarded engineering as a well-rewarded job rather than a vocation. However although most American and British studies do show that engineering students and engineers tend to focus their interests quite strongly (and not surprisingly) on technical matters, and although they offer some evidence of hard-worked students having slightly and temporarily restricted outlooks compared with some other students, their authors tended to play down the very *demotic* nature of most engineering tasks. By using the term demotic, we emphasize how most engineering involves working with a wide range of people in highly practical ways, with effects on the lives of a wide range of other people. The idea that engineers lack concern for people seems to be a rather spurious one, partly due to inadequate research design of the studies of engineers, and partly to the prejudices of some social scientists. It also seems possible that a popular stereotype of scientists as 'madmen in laboratories' might have contaminated their attitudes to engineers by association (Glover, 1973).

Some of these studies seemed to have elicited misleading opinions about matters in which engineers had little direct interest, opinions which were then wrongly labelled as 'values' (Taguiri, 1965–6). Support for this suggestion comes from two studies which asked engineers in industry about their political views and sources of job satisfaction. In one case, engineers seemed to place little importance on order and certainty: in no way were the typical respondents' socio-political values significantly 'authoritarian' or 'object-oriented' (Glover, 1973). In the second, similar results concerning values were obtained, and on sources of job satisfaction the gratifications which came from working with, being responsible for, and helping others were emphasized strongly (Bamber and Glover, 1975). Just because some engineers, because of differences in job content, are less directly concerned with socio-political issues than personnel managers, or less with ideas than scientific researchers, hardly makes them authoritarian. In fact their views seemed to be *more* democratic, less elitist than members of the other two groups in both of the studies.

Another factor which may have influenced conclusions drawn about the narrowing aspects of engineering in the earlier studies was the relatively low social and academic status of engineering in Britain, and to a smaller degree, the USA. Yet all specialist occupations have their own narrowing effects, and in fact engineers probably have more right to be called 'humanists' by virtue of their varied and often continual contact with people than many others, especially such bookworms as professors of literature or social science, and such bureaucrats as some personnel managers!

In settings in which engineers enjoy superior prospects and status to their British counterparts, one might expect their attitudes to seem more confident and liberal. A major West German study of over 25,000 engineers conducted some ten years ago seems to confirm this. Kogon (1976) found that West German engineers were neither uninterested nor uninvolved in politics and that their views were far from being 'right-wing, reactionary, or traditionalist'. Nearly a third ranked themselves as left-wing, and less than one in seven ranked themselves right-wing: the others were in between. Asked to choose from a list of political adjectives which described them best, 'the first choice was *anti-communist* closely followed by *liberal*; the third choice was *left–liberal*' and the 'proportion choosing *socialist* was more than double those choosing *conservative*'. They were not militaristic, deferential to established power, nor nationalistic, and they kept themselves well informed on current affairs. They were critical of abuses of big business power, and they were sympathetic to trade unions and to the system of employee participation in West German industry even though most did not belong to the former or benefit directly from the latter. Although hostile to welfare state 'spongers', 'worried about sexual freedom and ... . not opposed to occasional corporal punishment to children', they were strongly in favour of pupil participation in school administration because they felt it helped to develop democratic attitudes. They were also fully aware that the technically possible was not always socially desirable and they accepted 'responsibility for the uses to which engineering is put'. Lawrence (1977) used this evidence, and *lack* of evidence for the British stereotype, to suggest that British engineers may often be excluded from positions of economic, political and social leadership simply because they are *assumed* to have the wrong attitudes.

Engineering and the State

The term 'the state' refers to institutions of government and to individual makers and enforcers of laws and public administrative decisions. In Britain these include professional politicians, the monarchy, members of both Houses of Parliament, civil servants, judges, the police, local government officers, the armed forces, and the established church. Among the most powerful of these are professional politicians and senior civil servants. Private interest groups such as the Confederation of British Industries, the Trades Union Congress, and even the main political parties, some of whose members and officials are members of the previous groups, do not usually form part of the state as such, although there are times when they are effectively co-opted into it by elected governments. Here we are concerned primarily with the general nature of state power and activity and its varied and often quite complicated relationships with engineers and engineering. It is important to remember that neither engineers nor state institutions are independent of the wider societies to which they belong; and there is no such thing as a simple 1:1 relationship between engineering and the state, anywhere.

Social scientists have long debated the nature of state power in many settings, but a central theme concerns the degree to which dominant economic classes rule politically. Left-wing writers have usually argued that ruling classes who own and control productive property also control either directly or indirectly both the machinery of government and the ideas which shape everything else. Those with more liberal or right-wing views point to the complex division of labour in industrial societies, arguing that power is usually divided between various elites – managerial, educational, religious, cultural, military, political, administrative, and so on – which tend to be separate in terms of backgrounds, training, organization and values. Recruitment to them is felt to be at least partly based on merit, and the elites are thought to be interdependent, with no single one permanently dominant. Political elites are thought to be 'honest brokers', in complex, 'plural' democratic societies. The state's institutions are thought of as relatively weak servants of politicians and the people, not as forming the repressive apparatus of a simple, unified, economically and politically dominant class (Dahl, 1963; Miliband, 1969).

In Britain support can be found for both views. It would be ridiculous to suggest that a tightly-knit group of wealthy business people think and act in concert to control everything, but it is easy to show that wealth and the kinds of education that it helps to foster, and indeed buy, do greatly influence events (Stanworth and Giddens, 1974). Of course Britain is not unique in this way. Although the extension of the franchise has meant that wealthy people do not influence political decisions to the extent that they once did, political representatives still tend to be significantly wealthier than their electorates, as are senior executives of such non-political British state institutions as public corporations, the civil service, the armed forces, the Bank of England and the National Health Service. British academics and, to a smaller degree, senior civil servants, tend to come from humbler family and educational backgrounds than many of the others. Executives in private industry come from quite a wide range of backgrounds, but a privileged education remains a valuable if not an indispensable asset, and at the highest level of the largest firms executives are harder to distinguish from counterparts in other sectors, with whom they often enjoy family and social connections (Stanworth and Giddens, 1974; Leggatt, 1978).

Because motives and decisions cannot be observed, the existence or otherwise of a genuinely self-seeking ruling class is virtually impossible to prove, and it will no doubt continue to be endlessly debated. So will the chicken–and–egg issue of whether property itself is a source of power or whether political, legal and other forms of authority give control over property. Although powerful positions are very often occupied by wealthy people, that does not necessarily mean that such people will always act selfishly. Also they do not need to fill positions in government or administration in order to rule: they can do so by proxy, influence, ideas, by pulling strings, and so on. However, a ruling class may be said to exist insofar as inequalities of property, income, status and power persist over time, and as long as the same groups continue to profit accordingly.

A capitalist ruling class can also be said to exist insofar as a belief in, and the institution of, private property hold sway over such opposed beliefs and institutions as those of communism. Any system of beliefs can act as a kind of front behind which monopolies of force and the ability to manipulate material rewards are hidden. However, economic and technical development throw up new kinds of entrepreneur and skilled employee, all with a desire to influence decisions. Affluence and large-scale organization also lead to the growth of pressure groups which seek and exert political influence. All such changes tend to weaken the position of inherited wealth. One answer for the established powerful is to use education and the mass media to promote consensus so as to help people to believe for example in freedom, the monarchy, parliamentary government and peaceful evolution as opposed to bureaucracy, privileges given to experts, the dangers of 'materialism', 'political' forms of employee action against employers, and violent revolution. The influences of education and the media are easy to exaggerate: both tend to follow society in the first instance rather than vice versa. Nevertheless they often do have a great deal of subtle and sometimes not-so-subtle power and influence.

To shed light on the ruling-class versus competing-elites question in Britain, we need to know about the origins and development of social groupings, and about the degrees to which they are self-recruiting, self-sustaining and interdependent. We also need to allow the possibility that there is little or no truly centralized source or exercise of power. It is also reasonable to note that on many important matters, either no decisions are made at all due to indifference, ignorance, opposition or lack of resources, or that ineffectual decisions that are made are often simply window dressing or a reflection of impotence or an unfounded sense of self-importance. Or decisions 'happen to' people in the form of rationalizations after events which turned out unexpectedly well (cf. Mintzberg *et al.*, 1976). Certainly 'Britain is not ruled by a small self-conscious conspiratorial group of men' because in reality 'both decisions and non-decisions emerge from the interdependent actions of a number of men within large institutions, who may at times be in much competition with each other', with 'final outcome[s often] unintended by anyone', irrespective of the fact of significant inequality in the distribution of power (Urry, 1973).

Many commentators have noted how countries in which industrialization came about 'from below', due to the actions of independent entrepreneurs, have tended to develop multi-party political systems, whereas those in which industrialization was imposed from above have tended to become one-party states. Differences between the attitudes and behaviour of old landowning/ruling groups can very often account for this, along with differences in sources and rates of technical and economic change.

By most standards Britain's experience of industrialization was slow and mainly peaceful, and although it was far from typical in other very important ways, this has enhanced its attractiveness as an example of evolutionary change. British governments of the eighteenth century and after generally adopted a highly permissive stance towards growth of internal trade. The relevant decisions were part of a broadly liberal package of freedoms, and the effects of the tougher decisions were often exported, or exported themselves, to the Empire and elsewhere overseas. In several societies keen to catch up with Britain, industrialization was very much imposed from above by governments, usually against the wishes of landowners, peasants and religious leaders (Moore, 1967).

It is mistaken to think too glibly and simply of industrialization as representing some kind of ultimate goal or end-state to be achieved by the industrial and the supposedly 'developing' countries. Virtually all of the industrial countries share factory systems of production, market systems of distribution, an elaborate division of labour, and attempts to make education relevant to production (Feldman and Moore, 1965). But industrialization's benefits can permit the retention of traditional ways of acting and thinking, as for example with Britain and Sweden's retention of the monarchy and widespread religious observance in the USA. Industrialization does *not* sweep all before it. We should also remember that the economic 'bases' of industrial countries are often at least as varied as their cultural, political and social 'superstructures' (cf. Kerr *et al.*, 1973).

Dore (1973) suggested that pre-industrial history, exigencies and the nature of the industrialization process as such were three sources of the diversity of industrial societies. By exigencies he meant such relatively unique and powerful influences as wars, international relations, revolutions, the beliefs of elite groups, population size and density and its social and ethnic composition, and natural resources.

More specifically and with more direct relevance to present concerns, Sorge (1979) showed that the evolution of different systems of technical education and training in Britain, France and Germany could not be explained adequately without understanding the very different political histories of the three countries. He identified three broad philosophies concerning the role of the state and others in technical education and training. With the *individualistic* philosophy, training was the responsibility of employers and employees. The *association* view suggested that training was the responsibility of public but non-state bodies such as professional associations, craftsmen's trade unions, and chambers of commerce. The complete opposite to the individualistic philosophy was the *state school* view. Here training was the responsibility of governments who established and ran technical and other vocational universities, colleges and polytechnics.

Sorge explained how each of the three countries had evolved its own approach out of an originally common medieval heritage of the guild system (and of universities which had mainly vocational aims). The craft guilds had set standards of work and regulated the training of artisans. They were responsible for the flow of apprentices and their examinations. Thus they controlled the numbers employed in the various trades. This right was granted them by a prince, a state or a city, and the system was based on the association view.

In Britain and the USA, in contrast to France, Germany and several other major industrial countries like Japan and Russia, industrialization came from below. The power elite model of society seems more applicable to Britain and the USA than the ruling class one. In other countries ruling groups were more homogeneous and unified during state-imposed industrialization. This comparison involves a lot of simplification and the role of the state has since changed somewhat in most industrial countries. Nevertheless the comparison is broadly valid and it offers a sort of backcloth to understanding current relationships between the state and engineering in Britain, where a *laissez-faire* approach has largely operated.

Politics and Wealth Versus Engineering?

Is there a general tendency for political and financial forces to enhance or to reduce the power of engineers? Are engineers inevitably or endemically threatened by their own success? Does a society's achievement of wealth and power usually precede its decline by some kind of international leapfrogging process?

There does seem to be some merit in the idea that engineers represent those who live by their hands; and that financiers, politicians and social commentators live by their mouths. There is also sense in the idea that as nations and groups become wealthier, success breeds a tendency among their members to query the distribution of wealth along with complacency about its continued creation.

It is not therefore surprising that, in the affluent democracies, there has recently been a growing tendency to query the *logic* of economic growth. Industrialization would ultimately, it used to be felt, mean high living standards and considerable personal freedom for all. The power of inherited wealth would give way to that of the expert; subsequently experts would serve rather than rule the people. However experience showed that experts like engineers lacked a great deal of the power which had been attributed to them: most were still employees and servants of the powerful. Economic growth was often seen, on closer examination, to have undesirable side-effects and to owe more to greed than to reason.

To justify the status quo in many developed and other countries, powerful groups typically argue that achievement rather than inheritance is the main determinant of differences in rewards. In the West, property ownership is also regarded as a major source of social and political stability. It is recognized that the second idea is not completely compatible with the first one, but the contradiction is often papered over by pointing to the role of government in helping the weaker members of society, and to economic progress as a source of benefits for all. Competition, personal responsibility, and a broadly practical approach to life are also valued, often in ways which inhibit questioning of the nature of social relationships. Left-wing politicians normally praise the state's welfare and redistributive functions, whereas those on the right are more interested in the defence of individual rights, responsibilities and privileges.

When the creation of wealth is given more overt priority than its distribution, and especially when there is talk of 'the real economy' and 'real', tangible wealth as opposed to profits from services and financial manipulation, engineers are likely to be at their most influential. But they do not always fare so well when times are hard: in such periods such people as accountants and personnel specialists may fare better because they are more obviously trained to control and rationalize production and employment.

Bell (1978), mainly interested in the USA but also writing in ways which are quite relevant for Britain, argued that production, politics and self-expression all involve different values and behaviour in the West which were increasingly in conflict with each other. Economic life traditionally valued self-discipline, hierarchy and thrift. Politics emphasized participation in decision-making and shared citizenship rights. The cultural realm emphasized self-expression, increasingly in the face of fantasy, impulsiveness and self-indulgent hedonism and in an economic setting in which production was becoming simply a means to the end of consumption. Opposing Marx, Bell argued that culture is no longer (if it ever was) simply a product of economic power, insisting rather that it is an increasingly autonomous force directed against power and all established institutions. Renewed economic difficulties followed: the self-interest which has made us rich was no longer enlightened, having turned self-destructively on itself.

Bell argues how economic concerns focused increasingly on so-called 'information revolutions'. Wrongly, people thought that these were in the vanguard of progress. In these circumstances it was suggested that the cultural realm, dealing mainly in information, would be anti-industrial. At the same time businessmen would want to transcend business and to deal in clean abstractions and symbols of the kind taught on management courses. Because social status would depend increasingly on conspicuous consumption rather than on the effort to produce, the culture tended to 'institutionalize envy'. Many would-be 'liberationist' pressure groups would and had developed to accuse governments of being bureaucratic, inhuman and incompetent. Governments' tasks of deciding on priorities in the face of innumerable competing claims would become impossible as the weakened economic realm could no longer deliver the necessary goods.

Echoing Bacon and Eltis (1976) as well as Bell, but in the British context, Shanks (1971, 1978) also wrote of a runaway propensity to consume. For two decades after 1945 Britain experienced its fastest-ever rise in living standards, until economic difficulties surfaced and worsened into the 'stagflation' of the mid-1970s and after. Increasingly unrealistic expectations of the productive sector had been generated; the sector itself became more and more undisciplined and unable to satisfy them. Countries whose values emphasized social consensus, like Japan, Sweden and West Germany, had been better equipped to ride out the economic turbulence of the 1970s than such English-speaking industrial countries as Britain, Canada and the USA, with their traditions of selfish individualism, radicalism and consumerism. So if these arguments are valid – and they are more plausible than they are testable – engineers and engineering are not only threatened by their own success, but by the values prevalent in some of the environments in which their successes occur. More significantly however the arguments themselves are a little flawed by their reliance on an idealized version of the distinction between the productive and non-productive sectors of the economy. Nevertheless the issues raised are important and interesting.

Values concerning industry and trade and the aims and backgrounds of top economic and political decision-makers strongly influence the extent to which engineering is subjected to financial interests and constraints. The continued importance of personal wealth in the control of large British and American companies has been described by Fidler (1981), Francis (1980), Scott (1979, 1981) and Zeitlin (1974). In most of the industrial countries salaried managers including engineers are effectively subordinated to wealthy individuals and financial institutions. Personal wealth and share ownership, interlocking directorships and educational, family and social links amongst top job holders have all helped to produce a business elite of a relatively small number of large shareholders, financiers and capitalist entrepreneurs. Its character seems not to have changed fundamentally for decades, at least not in the Western countries which were victorious in the Second World War. In larger British firms good 'professional management' seems to mean the successful pursuit of profits, growth and return on investment rather than the technical quality of processes and products, or 'socially responsible' behaviour towards customers and employees (see for example Pahl and Winkler, 1974; and Poole *et al.*, 1981). However, in the final analysis pursuit of technical quality and the good management of people may result in more soundly based financial strength than will the pursuit of short-term profits.

Taking this last point further, Bellini (1981) argued that British financial interests were increasingly orientated, in a new 'feudal' fashion, to land ownership and to information sectors such as banking, insurance and finance, and overseas investment, as sources of income. Accelerating neglect of British manufacturing industry and the creation of new classes of 'serfs' were central features of this strategy. It does seem that the proportion and influence of senior British managers in manufacturing with financial and commercial backgrounds has increased since the 1950s at the expense of those with technical ones (Poole *et al.*, 1981, Chapter 2). Fidler (1981) argued that 'a high proportion [of top British job holders] have economics, commerce and accounting education, in which they are likely to have imbibed classical economic theory' with the idea of profit maximization at its core. Such people were often in favour of the socially-responsible 'soulful corporation', but only when they felt it was compatible with profit-seeking, economic security, survival and expansion. Many studies of middle managers, while noting their greater concern for their specialist positions and functions compared with their superiors, also show that they generally share the same background socio-political values, especially insofar as managerial authority is concerned.

Kaldor (1984) made a speech to the House of Lords, echoing Bellini by criticizing recent British readiness to allow indigenous manufacturing to run down in favour of services and of investment abroad. Much of the income from services recorded in balance of payments statistics consisted of interest and dividends received from abroad, not payments for services provided by the City of London. The latter consisted of only about £1 billion. Other services such as construction carried out abroad by British firms, royalties to authors and playwrights, and consultancy fees added about another £1.5 billion. Along with air and sea transport (£0.5 billion) and interest and dividends from foreign investment, such 'exported' services provided a total benefit to the balance of payments of only £5 billion, enough only to pay for less than a tenth of Britain's current imports. The economic contribution of services could only be seriously increased in a socially regressive way by generating jobs in domestic and other menial employment sectors. As ways of increasing wealth, many services were neither particularly important nor desirable, and it was only a defeatist and prejudiced stance towards manufacturing which made us look to them. Britain was not at a natural disadvantage in manufacturing; it had plenty of readily accessible sources of coal and iron in Britain and Scandinavia (as well as North Sea oil and gas). Japan's steel output had multiplied many times since 1945 while ours had increased by less than a quarter. Yet Japan had to import both coal and iron from distances of 10,000 miles or more.

Britain had refused to invest in capital goods, in shipbuilding and in all kinds of engineering, in spite of originally being better placed to do so than Japan or several European countries. Instead Britain had placed its faith in light industries and services. But without heavy industry Britain could neither expand the economy nor fight wars or deal with some other really major crisis for more than a few days or weeks. High costs and lack of technical competitiveness had been used by the British as excuses for further decline, as counsels of despair, instead of as reasons for investing in hardware and engineering education. The benefits of the North Sea were being spent on social security payments and on the selfish, narrow pursuit of investment income from abroad, to the short-term benefit of wealthy Britons but to the detriment of everyone else.

Finally, we can accept many of the points just noted about the rather selfish and irrational attitudes which manufacturing suffers from, while still agreeing with some 'environmentalist' criticisms of its effects, acknowledging that there can be limits, of people, energy and materials, to economic growth. A study by Cotgrove (1982) suggests that, while both optimism and pessimism about industry's effects are 'faiths and doctrines' in the last analysis, optimism is both more common and more popular with vested interests. Yet 'It would seem prudent, for the long-term interests of us all, to pay particular attention to the warnings of the [pessimists, who] . . . *could* be right'.

PART TWO

In the first part of this book we have articulated a number of issues surrounding engineers and engineering. Our purpose has been to describe the context within which engineering and engineers operate as well as laying out some of the basic assumptions which we as sociologists bring to bear on the issues. However sociology is not just a descriptive exercise, and the purpose of this book is not simply to describe the problems facing engineers. Sociology aims to explain social phenomena. To this end the reader should keep two things in mind. First, sociological explanation involves identifying relationships between phenomena. These relationships are variously associational, correlational and causal. Generally speaking the greater precision with which phenomena can be identified and defined, the greater the precision in identifying the nature of the relationships. Once such relationships can be delimited then the process of comprehending their nature can be attempted.

Second, in order to help illuminate relationships sociologists use a battery of theoretical or conceptual ideas as tools. We have suggested that *homo faber* and the division of labour are useful in understanding engineering. For the most part the rest of this book is concerned with an analysis of the division of labour as it applies to engineering. This is not because we wish to ignore *homo faber*, but rather we have made an assumption based on the idea of *homo faber*, i.e. that tool-making and -using is central to human life. An analysis of the division of labour helps us to understand the processes which push man the tool-maker and -user off centre stage in Britain.

The division of labour is not a new sociological idea, but one which, as we argued earlier and demonstrate below, goes back to the very roots of the discipline. The division of labour is a means of comprehending two things simultaneously about social and economic life. In the first place there are recurrent patterns in human conduct – both groups' and individuals' collective and isolated behaviour fits into patterns – not in a preordained way, but rather in the sense that, viewed in an aggregate perspective, human behaviour tends to be the playing out of regular sequences of actions. In the second place, human society and human societies are endlessly variable insofar as groups and individuals are differentiated from each other and ultimately no two individuals' actions are ever identical; human groups and institutions are in an endless state of flux and turmoil as millions upon millions of individual actions make up the rich fabric of human life. The division of labour both describes the differentiation between individuals and groups and the similarities and repetitiveness of human life. The division of labour refers to structures – the similarities in human conduct and people's experiences of those structures – and to processes, through which the structures and so on change.

The second part of our book identifies several important aspects of social structure and processes as worthy of particular attention in the context of engineers. These include education and training, aspects of the structure and process of work organization, and the system and process of collective organization outwith formal work roles (the trade unions and professional associations). We also address the experiences of engineers and others within these systems and processes. In Chapter 11 we bring the arguments together in a detailed examination of the division of labour itself as it applies to engineering. At that point we hope that we will have demonstrated that the key to understanding the nature of the problems we identified in Part One of the book can be more readily comprehended – and, perhaps, then acted upon – by using the theoretical framework which we have advanced.

6 Engineering Education

Introduction

Four themes dominate this chapter. First we explore the general relationships between education and engineering and in particular examine some of the confusions which have surrounded discussions about engineering education. Second, the historical legacies of *laissez-faire* and humanism as dominant values in respect of education in Britain are explored and their relationships to the current organization of engineering education are delimited. Finally, in the conclusion we examine the social–psychological consequences of these factors on individuals choosing to become engineers in the first place and undergoing engineering education itself.

Engineers, Education and Society

In post-war Britain there has been a long and controversial debate about national needs for 'scientific and technological' manpower. From time to time politicians, with some professors lending strong support, have been attracted by the idea that education should be encouraged and directed to produce specific numbers of people possessing particular types of qualification and skill. However others have argued that there is no systematic pattern, across countries, in the relationship between the education of workers and income per head or rates of growth of national income, either for economic systems as a whole, or for individual sectors of economic activity (see Gannicot and Blaug, 1969; Blaug, 1965; McCormick, 1979). This finding applied even more strongly to graduates than to other salaried and wage-earning employees. Major countries had fallen for the idea of manpower forecasting in the early 1960s but their forecasts had virtually all failed to predict what they had promised (OECD, 1970).

One point of view is that short-term forecasts of one, two or three years of the broad requirements for people with definite amounts of schooling are of some use, but that the more ambitious and specific types of forecasting are more or less either impossible or not feasible. This is because it is almost always possible to find substitutes for skills, by re-designing jobs, by mechanization, by training, by importing skills, or even by importing products.

Behind this criticism there also lurks the impossibility of knowing which tasks will need to be performed in the future, because of technical and market changes which can rarely be predicted (even in the fairly short term of the length of a degree course) and which very often occur outside the frontiers of the countries which they affect (Blaug, 1965).

Other questions about efforts to tie education and training to ideas about the content of jobs have been raised by Berg (1970) and Dore (1976). Studies reviewed by Berg for the USA (1970) showed that more highly and relevantly educated employees – in a society with a very diverse and complex education and training system – were by no means likely to be more able or productive than the less educated. While 'it would be foolish to deny that education is involved in the . . . capacity to produce goods and services' (Berg, 1970) the very complicated nature of education and of jobs meant that there were few clear links. A favoured minority in the USA generally had too much education whereas many other people did not have enough. Employers were guilty of naive 'cre-dentialism'; they raised broad entry requirements for jobs as educational levels rose, but this did not mean that the more educated were more able or more relevantly trained than others, although they were paid more. Similarly Dore (1976), in *The Diploma Disease*, attacked 'mere qualification earning'. He argued that schools and universities all over the world were allowing their 'sifting function', of sorting people into winners and losers, to dominate their ancient role of providing educa-tion. The tendencies he describes were worse in the developing countries than in the richer ones, but everywhere more and more people were labelled as failures, more and more people had imagination and curiosity squeezed out of them by 'education', as more and more schools and universities were being set up in order to 'widen opportunity'.

In his chapter on English education, Dore (1976: 22–3) described how the nature of civil engineering qualifications changed between 1897 and 1971. The transition was one from apprenticeship to mid-career quali-fication to post-career qualification, from on-the-job learning to part-time study to a full-time degree. Accountants', solicitors' and librarians' qualifications had also been travelling the same kind of route. Dore's worries about the usefulness of the trend relates to several debates about the best ways of educating and training engineers.

It has often been argued that the engineer-entrepreneurs and 'practical tinkerers' who created Britain's industries in the century or so after 1750 were conspicuously lacking in scientific training (see for example, Mathias, 1969). It is also often suggested that Britain's relative economic decline after 1870 or thereabouts was due to the lack of scientific training for the supposedly more sophisticated types of engineering which were introduced when chemical, electrical and other so-called science-based sectors were established. Certainly Britain, or more especially England, was slower to develop higher engineering education than were her main competitors (see for example, Landes, 1969; Gowing, 1978; Allen, 1979; Ahlström, 1982). However, a reading of these last-named sources, and also of the comments of Fores (1979), suggests that many of the points typically made about such education confuse several issues. Barnett (1972: 94–9) attacked British industry for its professional and technical ignorance and its faith in the 'practical man' and in the 'rule of thumb' in the nineteenth and twentieth centuries. Yet the simple idea that current or indeed any methods of manufacturing demand specifically 'professional' knowledge and skills or that they simply consist of the application of science as suggested by Mathias (1969), Landes (1969) and other economic historians, is a particularly British view (see for example, Fores and Glover, 1978b; Glover, 1978b; Finniston, 1980; and Hutton and Lawrence, 1981). Professions in the British or North American sense scarcely exist on the Continent of Europe; and the whole nature of the ways in which the French and Germans and other countries influenced by them produce engineers serves to deny the whole Anglo-Saxon idea – which is apparently validated by various BBC and other TV and radio programmes, the concerns and titles of Imperial (and other) College(s) of Science and Technology and of the Science Museum – that engineering is a branch of science. In fact the issues confused in writings on engineering and engineering education concern the nature of engineering, the way in which it has developed since the eighteenth century, the requirements of engineering education, and the status and attractiveness of engineering careers.

On the nature of engineering and on the ways in which it has changed, confusion arises out of the belief of many academic observers, especially in countries like Britain in which its status is not high, that engineering increasingly consists, as it becomes more difficult, of the use of scientific knowledge and the application of scientific principles (Swinnerton-Dyer, 1982). In this way, it is felt, it merits rather pretentious descriptions like 'high technology' which implicitly stress its knowledge content and purely intellectual difficulty. Yet as Fores (1972, 1979) has argued, the essence of engineering has never changed and it has never been, nor can it ever be, part of science or an '-ology' of any kind.

This is primarily because the defining outputs of engineering are normally useful three-dimensional objects produced with some commercial or practical aim in mind, whereas those of science take a two-dimensional paper form (articles in learned journals) and are produced in the pursuit of truth. As we have pointed out earlier, engineering does, of course, use scientific knowledge, and technical knowledge of many types concerning markets, money and people as well as hardware, but so do most other human activities.

The confusion apparent in writings on engineering education are at least a century old and are partly due to the undoubtedly increasing scientific background to a great deal of engineering work. In a country like Britain in which the disinterested pursuit of knowledge for its own sake has been given more prestige than production for practical and commercial reasons, engineers have often given way to the temptation to shelter under the wing of science in education. Not only has Britain produced more science graduates relative to graduates in engineering than is the case in France, Germany and Sweden, for example (Prais, 1981b; Ahlström, 1982), but the emphasis of engineering education in Britain has been notably less practical and more scientific than in several major competitor countries (Finniston, 1980: Appendix E). Thus there are both status and practical reasons for the British tendency to accept the misleading and pretentious descriptions of engineering as applied science or technology.

It is also clear from Continental and other cross-cultural comparisons (Glover, 1978a; Swords-Isherwood, 1979; Finniston, 1980; Hutton and Lawrence, 1981; Ahlström, 1982) that engineering education is very often broader in scope outside Britain and that engineering careers attract a higher proportion of the most able members of each generation. In countries in which engineers are regarded highly, in which they have attended prestigious educational institutions and in which they enjoy good promotion opportunities, it is more likely that they will be listened to. This does not mean that their education is less scientifically sophisticated, rather that they will be less likely to try to dignify their work with scientific, or professional, or managerial status, because their status *as engineers* is already so high. Much of the British confusion about engineering education is simply a function of the relatively low social status of engineering itself.

All these last points relate to the previous ones about the faith of many politicians and others in education as an economic panacea, and to the 'credentialism' of some employers. The latter can take two forms. On the Continent the level and specific nature of a qualification is seen as directly relevant for slotting people into jobs, whereas in Britain most employers have tended only to assume that young graduates are more intelligent than young non-graduates. A broad view of the links between education and the economic performance of nations would not adopt the rather innocent idea that the study of hard and relevant subjects forms a large part of what is needed. The British example seems after all to disprove it in some important respects: Britain has a superb record in the study of such subjects if the performance of her natural scientists is anything to go by, and a generally poor record in making things. What does seem to count is sustained and detailed commitment, probably at all levels, but especially at the highest ones, to manufacturing and other forms of wealth creation.

In Britain there has been a historical commitment to the idea that the teaching of useful skills should be left to individual employers, helped by associations formed by members of occupations, such as professional bodies and the craft type of trade union. The French in contrast have long believed that training, at all levels, should be defined by a high measure of state involvement through the provision and supervision of vocationally orientated schools (Fores and Glover, 1976a). This view has affected all kinds of useful education and training in France. It has been predominant in Germany at the managerial level since the nineteenth century, although the involvement of the state at the lower levels is much weaker there, with something like a mixture of the British type of technical college and apprenticeship systems, albeit much more comprehensive and rigorous than is the case in Britain (Sorge, 1978a). The managerial-level approach to education and training for manufacturing and commerce which Germany produced in the nineteenth century drew on and modified French experience. It has produced able, broadly educated engineers with sector-specific skills and has influenced many European and other countries. Interestingly Scottish education, which has long reflected Continental practice in some respects, had an important (if indirect) effect on Japanese development. As Dore (1976:42) relates, 'a 25 year old Scot who was scornful of the engineering education he had received in Glasgow and was full of ideas about new methods of training he derived from Zurich and elsewhere' picked a team of eight fellow-Scots who founded the prestigious Imperial College of Engineering in Tokyo in 1873. This was 'the decisive act' in the establishment of Japanese engineering education.

Sorge describes how international differences in systems for producing manpower are best explained by 'Political factors in a country's social history'. British leaders had little liking for or understanding of manufacturing and they felt that economic competition would, and ought to, make enterprises take care of training with the help of occupational associations. The French, especially at the time of their Revolution, felt that occupational associations were harmful to the public welfare, and that the republic rather than associations of some of those whom it represented, should take care of all kinds of education and training. The medieval guilds' rights to provide apprenticeships were abolished in France in 1791, and all associations which aimed to regulate wages and conditions of work and entry into trades were forbidden by law shortly afterwards. In fact the French had shown their faith in the centralization of state power for at least two generations before the Revolution, by setting up engineering schools such as the *Ecole des Ponts et Chaussees* in 1747 and the *Ecole des Mines* four years later, as well as by stimulating and regulating internal manufacturing and commerce for at least two centuries before that. It was not the setting up of the schools of engineering education as such which gave French engineering its prestige. The occupations of civil and mining engineering existed before the establishment of the schools. The occupations themselves were prestigious from the first and they were the main force behind the setting up of the schools.

The old political privileges of the medieval guilds were also cancelled and trade was liberalized in most of the German states at the beginning of the nineteenth century. Like France, Germany had a tradition of absolutist rule, and this was not completely replaced by parliamentary democracy and political liberalism until the twentieth century. The guilds were allowed to practise vocational functions much longer in Germany. However the French type of higher technical education was copied in Germany from the first half of the nineteenth century, while lower-level types of vocational education and training were also developed but with more freedom from the state, with more local roots and direct employer involvement. The old guild tradition along with other relevant local powers persisted in Germany rather than in France because Germany never had such a centralized state as France, even in the Nazi years between 1933 and 1945. Germany only became a nation state as late as 1871; West Germany today is a federal republic made up of states, whereas the French provinces have traditionally been very strongly dominated by Paris in 'the most centralised nation on earth' (Ardagh, 1973).

Many of the problems associated with unravelling the links between engineering education and industry are clarified if the fact is grasped that both engineering education and industry reflect the same cultural values of British society. British industry, particularly manufacturing industry, is made up of a great many advisers and relatively few doers. The education system reflects that division too in terms of course provision. In British industry the non-technical types tend to be more powerful than technical people both inside and outside manufacturing and industrial units. The education system similarly reflects this pecking order. From university senates to secondary-school staffrooms engineering professors and technical teachers tend to be less well placed in terms of strategic decision-making than colleagues in other disciplines (excluding perhaps only the social sciences). Within industry there is a great deal of specialization outside production. In terms of the development of curricula in schools, colleges and universities the same applies. In the next section we explore in detail the ways in which these social and cultural values have affected British education.

Engineering Education and Training: an Overview

British attitudes to education are in general characterized by two features. First a set of values has developed which give relatively greater prestige to those subjects which stress intellectual rather than hand skills. Second the British state has traditionally maintained wherever possible a *laissez-faire* policy (a policy of leaving well alone) in respect of educational provision. The legacy of *laissez-faire* as far as education is concerned is that England and Wales have never had a national educational system (unlike Scotland). In the late twentieth century it remains a mixture of public provision, state subsidized indirect control and semi-private initiative. Both the preference for intellectual skills and *laissez-faire* are part of broader British social cultural values which had become firmly established in the public mind by the mid-nineteenth century (although their importance and influence is much older than that). We suggest that the preference for the intellectual rather than practical developed against a background in which British supremacy had been relatively easily gained; an Empire, an industrial base and a fairly peaceful internal political system produced self-confidence.

The fact that Britain's advantages developed because of the fortuitous existence of plentiful natural resources, a highly efficient navy (where at sea if not in the Admiralty doing rather than thinking was highly valued) and competitor nations who were in a state of internal political turmoil was easily overlooked and conveniently forgotten. Self-confidence became over-confidence and a mistaken disdain for the very practical and technical skills which had capitalized on Britain's economic supremacy and political stability in the first place. *Laissez-faire*, in our view, could be thought of as an after-the-event justification for something that happened largely by itself. Economic growth and the development of industry began without state involvement. *Laissez-faire* became both an explanation – growth occurs because of the effects of market forces – and a rationalization for continued non-intervention. It fitted well with other British values and preferences (which may also have been part delusion) for sturdy independence and freedom of action. In education the consequences has not been independence so much as lack of direction.

Victorian attitudes to education encapsulated both the preference for intellectual skills and *laissez-faire*. Two sets of arguments dominated Victorian discussions about education. The arguments were first between radical reformers who believed in universal education against those who saw education as a privilege which would be vulgarized by extension to the masses as well as being dangerously revolutionary. What both sides in this debate shared was a belief that education was an end in itself, whose purpose was to humanize and civilize. The second argument was between the Anglican Church and the Dissenting religious groups. The Anglicans demanded a religious element in education over which they would have absolute control so, as they saw it, they could defend the nation from the abandonment of religious faith. The Dissenters on the other hand wanted to teach their own religious beliefs and argued for the right to do so in the name of religious freedom. Although the participants in both arguments may have been well-intentioned, their concentration on argument about principles was at the expense of education and vocational provision. The English and Welsh system therefore developed on an *ad hoc* basis, and it has continued to do so to the present. The vigour with which the arguments were pursued arrested the development of a national educational system and allowed economic *laissez-faire* principles to go largely unchallenged (Woodward, 1962; Ensor, 1946).

It is important to distinguish the English system from the Scottish system. Scottish education is a national system to a much greater extent than English education is. There is only one examining body for all school subjects, for example. Also the Scottish Education Department (SED) has exerted a much greater degree of control and direction over Scottish education than the Department of Education and Science has in England and Wales. A national system of elementary education has existed in Scotland since the beginning of the seventeenth century. Scottish universities, and more recently the Scottish Central Institutions (which provide non-university degree education), have generally offered wider access to higher education and served local needs to a much greater degree and for far longer than their English and Welsh counterparts (Wojtas, 1984; Hunter, 1972).

Universal and compulsory elementary education was only finally established in England in 1902. Elementary education, all that the vast majority received until after the Second World War, was designed to instil discipline and the skills of reading, writing and arithmetic. Secondary, grammar-school education was modelled on the principles of intellectual accomplishments and sport, not on hand skills. The reformed English public schools like Rugby, Eton and Harrow had self-consciously developed intellectual and sporting accomplishments which were then transplanted virtually wholesale into the new or reformed local grammar schools. Throughout the twentieth century the legacy of these accomplishments as the basis for education and the state's *laissez-faire* principles have continued to shape British education – particularly British technical education.

This non-intervention has thus been the hallmark of the attitudes of successive governments to engineering education. There was little effective concern from the state with the rapidly changing technical base of industry in the nineteenth century. Indeed it has often been noted that the engineering innovations that were an integral feature of the transition from an agricultural–rural economy to an industrial–urban one took place entirely without any systematic national educational provision in technical and engineering subjects. Given the scale of the developments in spinning and weaving, metal working, canal and railway construction and mining, the bedrock of British industrialization, this is fairly surprising. What is perhaps more puzzling is that even once the pay-off in terms of wealth and prosperity and the general raising of living standards accompanying the transition to an industrial economy became apparent, the benefits of developing technical education seem only to have been perceived, and then imperfectly, by very few people (Albu, 1979; Ahlström, 1982).

It was not until 1884 that a Royal Commission investigated the matter. It recommended that drawing lessons in schools be extended, and that woodwork and metalwork be encouraged. In 1889 the Technical Instruction Act was passed authorizing the levy of a penny rate by local authorities to aid technical instruction under the supervision of the Science and Arts Department. Under the provision of this Act the instruction provided tended to be scientific and intellectual rather than technical and manual. The existing elementary schools were unaffected by its provisions and consequently the numbers trained were small.

In 1884 Central Colleges were founded by the City and Guilds Institute of London at South Kensington for engineering and related subjects. In 1907 these were eventually merged into what became Imperial College of Science and Technology (part of London University). However, even though instruction was then available at degree level the bulk of people attending Imperial College did so on a part-time basis. At the beginning of the twentieth century the majority of engineers who received any formal training did so through an apprenticeship system. For example, until 1897, when the

Institution of Civil Engineers introduced a three-tier range of qualifications, there was no examination of any kind for civil engineers. In
the first decades of the twentieth century the majority of entrants to
civil engineering qualified by part-time study while serving an
apprenticeship. Only by 1935 did the number of graduate entrants in
civil engineering exceed the number of mid-career qualifiers. In fact in
all branches of engineering, including mechanical and electrical, the
part-time route to qualification remained very significant in Britain
well into the 1960s and it is still not insignificant.

Much of the history of the provision of technical education in the
United Kingdom under the *laissez-faire* system has been one of
'academic drift', the process whereby once established, a vocational
educational institution changes its character towards traditional
academic directions (Burgess and Pratt, 1970). In Britain institutions
were able to do this because of the lack of a firm central government
policy for technical education. At local level industrialists, manufacturers and other local bigwigs, although often present on boards of
governors of colleges, polytechnics and universities, have been in a
minority or have not had effective influence on policy. Senior
academics have oriented themselves to their peers in other institutions. Local roots have been shallow for many institutions which have
become gradually cosmopolitan rather than local in character
(Gouldner, 1957–8). Thus many colleges associated with engineering
and technical subjects, once established, aped those very institutions
which embodied the values of non-engineering subjects. Some of the
oldest centres of technical education, the central colleges and the
mechanics institutes, early on turned themselves into constituent
parts of London University or full universities in their own right.
After the Second World War the colleges of advanced technology in
Birmingham and Salford, for example, became technological universities which quickly became (apart from their provision of some
sandwich courses) virtually indistinguishable from the older civic
universities like those in Manchester and Sheffield. And even now the
polytechnics which partly took over the mantle of vocationally-based
education from the colleges of advanced technology tend increasingly
to concentrate on higher level degree work and postgraduate research
at the expense of HND/HNC-type teaching. The only real exceptions to this have been the central institutions in Scotland which
because of direct control by the Scottish Education Department have
retained their technical and vocational orientation and have not
drifted towards the traditional university model as have the English
and Welsh polytechnics (Wojtas, 1984).

The dominant view about the relationship between education and training in the United Kingdom has been to keep them quite separate. This is another legacy of *laissez-faire*. Education – implying formal theoretical learning – has been regarded as the province of academic institutions. Training – on-the-job practical learning – has largely been seen as the responsibility of employers. With the state's *laissez-faire* attitude this system has continued right up to the present. The separation of education and training has also been encouraged by academic drift because as the technological universities and poly-technics became more and more like traditional universities it was very often convenient for them to maintain or extend a discreet distance from the practical side of things, thus consolidating the separation.

The effects of academic drift are not confined to the institutions but have also affected courses. The pattern in recent years has been for an absolute increase in the sheer numbers of people graduating in engineering subjects. The problem, however, has been that this has tended to be at the expense of technician engineers. With the rise in the number of degrees in engineering there has been a decline in the full-time HND route and an even sharper fall in the numbers coming into engineering through the part-time HNC route. The result has been that the technician level is being starved of its best candidates as they are tempted away into the more rarified atmosphere of degree-level studies and the promise of full Chartered Engineer status, which used to be obtainable by HNC and HND holders. At the same time engineer-ing education at degree level tended until the late 1970s to become increasingly theoretical and academically specialized rather than practical, and concerned with research and development rather than pro-duction and the commercial and financial functions in manufacturing.

The history of academic drift is not remarkable. It is perfectly understandable given the high value placed on non-technical educa-tion in Britain and the non-interventionist stance taken by successive governments on technical and other education since the nineteenth century. In fact it is ironic that the governments which have been the most committed to policies of free enterprise and *laissez-faire* in this century (Mrs Thatcher's administration) should also have been the ones to have taken the most interventionist stance in education. This is seen, in particular, in directives to the UGC on sizes of courses, in the various initiatives that have emanated from the DES and the SED on information technology, in the attempt to introduce evaluations of school teachers, in the proposed reform of CNAA and in the granting of a Royal Charter to Britain's only private university.

The legacy left by the high value placed on intellectual rather than manual accomplishments has been profound. In Britain engineering does not have nor ever has had the academic glamour of the arts or the natural sciences. Engineering and other technical subjects have had and continue to have to compete on unequal terms with the more prestigious subjects. The result has been that the most able school children have been directed away from most practical subjects throughout their school careers. This has happened both directly because certain options were closed to the most able children and indirectly as school children often received advice in schools which indicated that a place at university studying a pure discipline should be the ultimate aim of all able school-leavers (Jamieson and Lightfoot, 1982; Tubman and Lewis, 1979).

The consequence is that engineering has generally tended to attract a greater proportion of relative underachievers as well as the poorer performers in the school subjects of mathematics and English. The range of ability of engineering undergraduates is wide with a long tail of lower academic ability as measured by GCE A level or Scottish Highers results. The problem is worsened by the fact that the average university or polytechnic department (in any subject, not just engineering) is not geared to teach across a wide band of ability. Most departments' *raison d'être* has been to teach high-fliers. The resulting high failure rates in the early years of many engineering courses, and the consequent lowering of morale and motivation, serves to confirm the lowly status of engineers, engineering students and their teachers in the pecking order both within educational institutions and in wider society (Finniston, 1980). None of this is to deny that there are not some exceptions. For example, some of the most prestigious departments and faculties of engineering have attracted school-leavers who have achieved more while at school than their counterparts in associated science departments and faculties. There are many first-rate engineering lecturers and professors with relevant and recent industrial experience and contacts. But the general message of our argument stands.

Indeed not only have engineering departments not attracted enough high quality students, but also there is evidence that neither have they attracted sufficient numbers of students vis-à-vis those entering other subjects. For example at the start of academic session 1982–3 there were 9,342 students embarking on one sort of engineering course or another in Britain. The equivalent figure for Japan was about five times greater. The 9,342 British figure compares with 17,896 starting social science subjects and 16,367 starting arts subjects. Of the 9,342 just 155 were beginning production engineering courses and 368 aeronautical engineering courses compared with 376 starting archaeology, 1,274 starting sociology and 635 starting classics. The ingrained opposition to certain subjects in British academic institutions has ensured that the proportion of university (and polytechnic) natural science and arts graduates to engineering graduates is much higher in Britain than in most other industrial countries.

We have argued of two traditions that one – concerning the nature of education – should be an end in itself and aimed at civilizing and humanizing people while illogically neglecting and denigrating the skills of making and doing, and the other – concerning a non-interventionist stance on technical education – have led to the particular development of education in general and engineering education in particular in Britain. To bring the picture up to date we examine the current routes into engineering. We argue that the two traditions still have a continuing and persuasive influence even in the most recent reforms proposed by and for the engineering bodies.

In the early 1980s several hundred degree courses in engineering were offered at 48 British universities, 28 polytechnics and 12 other institutions (in particular the Scottish central institutions and certain colleges of higher education south of the border). The Council for National Academic Awards (CNAA) validated approximately one-third of all degrees in engineering. The Technical Education Council and its Scottish counterpart (TEC and SCOTEC) validated sub-degree courses in mechanical and production engineering and in electronics and telecommunications (Finniston, 1984).

The distinction between degree and sub-degree level courses reflects some of the divisions within the engineering occupation or profession itself. More generally, of course, the spectrum of engineering is so wide because great diversity is needed in developing competence and creativity. Engineering education and training is divided into three tiers, chartered or professional, technician and craftsman levels. In recent years the professional chartered tier has made efforts to adopt an all-graduate mode of entry. In England and Wales a three-year full-time or a four-year sandwich degree, and in Scotland a four-year full-time or a five-year sandwich degree are therefore now the normal entry routes for aspiring chartered engineers.

Until 1985, in order to obtain a qualification as a chartered engineer (C. Eng.) after graduation it was necessary to be elected to corporate membership of one of the sixteen chartered professional engineering institutions which were the constituent members of the Council of Engineering Institutions (CEI) or its affiliates (see the first table in Chapter 10 for a list of the institutions). The Engineering Council (established in 1982) was given the difficult task of codifying and raising standards and of tackling wider issues which the CEI had not been allowed by its constituent Institutions to address. In various proposals presented in 1983 the Engineering Council aimed to provide a framework for the future development of engineering education and training.

The system of registration changed in 1985. A chartered engineer must now be at least 25 years of age and a corporate member of a chartered engineering institution which is nominated by the Engineering Council. There are also provisions for individuals engaged in engineering to become chartered engineers where no recognized chartered engineering institution for them exists. Candidates for chartered status must pass an examination in the principles of engineering set by the Board of Engineers' Registration. Alternatively they must have passed another examination or test which is acceptable to the board as being at least equivalent to a university degree at honours standard in an engineering or closely related subject. The vast majority enter by this route. The candidate wishing to gain chartered engineering status must also have undergone formal training in engineering or have held a position recognized (by the Board of Engineers' Registration) to be a position which has effectively provided such training, and must have at least two years' experience in a position of responsibility suitable for a professional engineer. These are general requirements and some of the chartered engineering institutions have additional ones. From 1984 onwards, a new structure for engineering education began to evolve and to be implemented. Engineering degrees will now increasingly have a two-tier structure consisting of 'enhanced' degrees and 'enhanced and extended' degrees. Both types of course are generally adopting an integrated approach to theoretical and practical teaching so that they are more closely related to the needs of industry. Enhanced courses aim to provide the route for the main body of chartered engineers. The enhanced courses will last for three years (full-time) in England and Wales and four years (full-time) in Scotland with additional time where a sandwich structure is adopted. These degrees will include 'engineering applications' and design as an integral part of the total course and lead to the award of the B.Eng. (rather than B.Sc) degree at honours level.

Extended courses are being provided for some of the most able potential high-fliers. It is intended that the numbers on extended courses should not exceed 20 per cent of the total number of engineering students in unviersities and 10 per cent in the public sector polytechnics and colleges. There is, of course, a definite element of British academic traditionalism, if not elitism, in this arrangement. These extended courses last four years full-time or the equivalent on a sandwich basis. In most cases all students enter the enhanced course and some transfer to the longer course on the basis of ability and inclination. The purpose of the extended course is to provide for the study of engineering in greater breadth, or to provide an engineering education in a range of engineering disciplines by being multi-disciplinary, and/or to incorporate business skills in finance, marketing, costing and management as part of the curriculum. The extended courses usually include a major design project and lead to the award of an M.Eng. degree at honours level.

Along with the new degree structure there is a new three-stage system of registration for chartered engineers. The first is the academic stage which is attained on completion of the B.Eng. or M.Eng. course. The second is the training stage of registration lasting about 24 months, the purpose of which is the acquisition of knowledge and skills. The third stage for full chartered status follows after another two years' experience in a 'professional' capacity. For persons whose engineering degrees are not accredited by the Board of Engineers' Registration or who achieve only an ordinary or unclassified degree, or for those whose main subject was perhaps mathematics or physics, rather different arrangements apply. These candidates must take additional courses before the first stage of academic registration is achieved.

The second main route into engineering, providing the second tier of the profession, is that of the Technician Engineer (T.Eng.). For entry to this status group until 1985 an individual had to be at least 23 years of age, to have had an academic qualification of a standard which was not lower than HNC or a City and Guilds full technological certificate and to have had a minimum of five years' engineering experience of which two years must have been devoted to practical training. From 1985 registration as T.Eng. involves a three-stage route. First there is academic registration on completion of HNC/HC or HND/HD. There then follows one year's training for HNC/HC holders and two years' for HND/HD holders to the training stage of registration. Full registration occurs after three years' working experience. There is also the possibility of transfer to C.Eng. status on an individual basis for T.Eng. holders.

The third and lowest tier of the profession is that of Engineering Technician (Eng.Tech.). Until 1984 this route of entry required that candidates were at least 21, had an academic qualification of a standard not lower than an ONC or a City and Guilds Part II final technician certificate, and a minimum of three years' engineering experience of which two must have been devoted to practical training. Under the current system there is again a three-stage process of registration. The first stage of academic registration is achieved on completion of ONC/C or OND/D. The second stage involves one year's training for ONC/C holders and two years for OND/D holders to the training stage of registration. The third stage is achieved after two further years of practical experience to full Eng. Tech. status. In addition an Eng. Tech. registration can be converted to that of T. Eng. with additional academic study (for full details see Levy, 1983). The purpose of the *totality* of the Engineering Council policy is to provide ladders and bridges for all who wish to take advantage of them and have the ability to do so.

In 1984 and subsequently the Engineering Council has proposed that more resources be made available to teach engineering and to fund the training of engineers in industry. These proposals are valuable for two reasons. First, they attempt to bring some order to a previously very messy structure. Second, they go at least part of the way towards redressing the balance in engineering education away from cognitive activity towards high-level practical and commercial skills and to making it more broadly based.

However the two traditions which have had such a profound effect on engineering education for the last hundred years will still probably continue to exert a pervasive influence. The *laissez-faire* tradition continues insofar as the state still does not take a direct role in guiding the direction of engineering education. Instead education remains the responsibility of a semi-private institution – the Engineering Council. A more formal and tight qualifications structure – desirable as that may be in itself – does not, and will not necessarily, remedy the problem of engineering's marginal status in political, economic and academic life in the United Kingdom. Indeed there is a danger that if the Engineering Council does not exercise very strong influence and leadership, the process of academic drift will continue with universities and polytechnics competing for the upper end of the market while the OND/ONC and HND/HNC (and HD/HC) level will be left to other less favoured institutions and departments. However it is fair to say that the Engineering Council recognizes such dangers and self-consciously pursues a policy aimed at changing all relevant attitudes and practices.

The tradition of stressing purely intellectual accomplishments characteristic of such British education is not obviously present in the recent changes, with sound practical training and integrated courses apparently being the order of the day. However the new system still maintains the education–training dichotomy with educational institutions responsible for the academic and intellectual side and employers responsible for training. Calls for integration are much more easily made than executed, given the complexity of engineering and the diverse interests likely to be affected, and in the end the new system may not do much to alter the fundamental division between the nature and location of education and training. In civil engineering for example (and sometimes elsewhere) training continues to be end-on rather than integrated. Current and future changes are and will remain complex and fascinating to say the least. The extent to which particular 'managerial' components of degrees adopt sector-specific or generalist engineering philosophies will be of particular interest. (We prefer the former). A flavour of elitism and of privilege is implied in the extended degree courses for M. Eng. candidates. While there may be nothing objectionable to elitism, the fact that the M. Eng. route may become a significant source of researchers and teachers of engineering, rather than as a route to industry, must be a significant worry. M. Eng. degrees are designed to include aspects of law, economics, management and so on, however, and the Ph.D. route will remain for potential academics and researchers. Yet even so, many Continental European 'doctor engineers' work in industry (including production) and Britain has a long way indeed to go before its industries become as well staffed with engineers as Continental countries' are. It is also very possible that by raising the top end of the qualification system, the HD/HC route will become further de-valued. It also implies that certain institutions will be regarded as suitable to teach M. Eng. courses while others will not. This may do nothing more than reinforce the existing educational pecking order, and reflect a set of values which have been associated with so much harm to engineering education and national economic performance in the past.

Our feeling is that the new proposals will not alter things funda-mentally because they mainly tackle symptoms rather than causes. However, as we have argued, those fundamental causes rest on a historical legacy and set of values in British society which are so deeply entrenched that to change them would require more than tinkering with a system of qualifications, a long-overdue, sensible and neces-sary, but not a sufficient, condition for improvement.

The Individual's Choice of and Subsequent Career in Engineering

Almost all commentators agree that there are two major features associated with the British education system and its treatment of and attitude towards engineering. The first is that engineering in industry does not attract sufficient first-choice applicants. In other words the sort of person who opts for an engineering course or a career in engineering in industry has not always made a particularly positive or personally advantageous career choice. The choice is sometimes made because other higher prestige opinions, for example ones in natural scientific research, are closed. The second is that where engineering is a positive choice it is because the applicant either has low aspirations in the first place or has his aspirations lowered by virtue of having made the choice.

Three different explanations have been advanced to explain why engineering is rarely a positive career choice. The first explanation concentrates on the school system and is concerned with the prestige of engineering therein. It is argued that the British education system with its preference for intellectual pursuits has discouraged the provision in schools and colleges of technical courses and made such courses where they are provided unattractive to students. This means that schools neither serve a technically sophisticated society and nor do they, because of their curriculum specialization, produce people of sufficiently wide education to become good engineers. Because of specialization the facilities for studying technical and practical things are denied to a significant proportion of people. In many schools the brightest pupils cannot take woodwork, metalwork, technical drawing or domestic science, for example, while those specializing in science in England and Wales are often denied the opportunity to study languages or history. The highest value in school education is placed on intellectual activity, narrowly defined, at the expense of other things. Among sixth-formers studying science subjects there is both a strong preference and positive encouragement to the brightest to aim for natural science degree courses rather than engineering ones. All this contributes disastrously to the generally unfavourable attitudes which children have been found to have towards industry resulting in too few of the ablest young people choosing careers in industry in general or engineering in particular. Other career alternatives are invariably considered first (Jamieson and Lightfoot, 1982; Smith, 1977; Tubman and Lewis, 1979).

The second explanation of engineering as a generally second-best choice concentrates on degree-level education. There is evidence that even for those students who make a positive decision in favour of an engineering course, a career in private sector manufacturing industry is not the option which many of these people subsequently want to take up. There are also data which show that students on engineering degree courses can easily be seduced to careers in research and development rather than production. University and college life tends to be organized in such a way that for some students a career in academic research and teaching appears a very attractive not to say gentlemanly way of spending one's time. This is reinforced by an ethic strongly held in universities (and schools) which is still transmitted to students, that the pursuit of truth through science or reflection is something much more worthwhile than making and organizing things for everyday use and commercial benefit. The effect is that even in engineering subjects the most highly qualified graduates tend to shun careers in manufacturing industry. The same process even continues to work on those who do choose such careers. Questionnaire surveys of newly graduating engineering job-seekers in industry have shown there was a tendency to evaluate future employers on the basis of whether opportunities for research did or did not exist. 'Research' often meant interesting technical work, for many such people, who seemed to have little concern for engineering's important commercial, financial and social aspects. At the same time job-seekers seemed to want to avoid as far as possible the production and sales functions (Cotgrove and Box, 1970; Herriot, Ecob and Hutchison, 1980; Smithers, 1969; Taylor, 1979).

The third explanation concerns engineering employers. It is suggested that many employers have failed to present themselves in a way that is attractive to prospective engineers. It is also suggested that they have failed to become involved in course design and teaching, while at the same time being rather suspicious of engineering graduates in particular and of education as opposed to training. Given the prevailing attitudes in educational institutions the employers' attitudes seem perfectly understandable (Herriot and Rothwell, 1981; Glover and Herriot, 1982).

We now consider individuals who have made a positive choice to do engineering, sometimes however with low career aspirations. Factors associated with this group have also been studied. Again three types of explanation have been advanced: the first is concerned with general attitudes, the second with types of mental aptitude, and the third with social background.

First, five general attitudes have been found which are conducive to a person wanting to become an engineer. These are: interests in technical things through pastimes or hobbies; the anticipation of a rewarding career in engineering; motivation through the study of engineering-related subjects at school; the advice of relatives and friends; and the advice and influence of teachers and careers officers. These last two factors were recognized as less important than the first three. What is striking about these factors is that given the structure of school education and its value system, the chances of even three out of the five factors being present together is statistically remote. It can be argued then that chance is important regarding the numbers of people who do make engineering a positive choice subject (Glover, 1973; Smithers, 1969).

The second approach has been to look at mental aptitudes. It has been found (by Hudson) that schoolboys destined for the undergraduate study of scientific and engineering subjects tend to score highly on intelligence tests, to have high numeracy and diagrammatic skills, to be able to work at high levels of accuracy and to have practical or mechanical interests. This cluster of aptitudes is termed convergent thinking and is contrasted with divergent thinking which was observed among potential arts graduates. Divergent thinking was characterized by relatively low scores on intelligence tests, high verbal rather than numerical or diagrammatic skills, low levels of accuracy in working, and cultural rather than mechanical interests. The sting in the tail of this argument is the fact that Hudson believes that subjects like engineering, which require problem-solving abilities and the capacity to see old problems in new ways, are not the types of skill which convergent thinkers are best at. In other words many of those students most likely to go into engineering, those trained in science subjects, were actually neither psychologically suited for it nor receiving the school education most appropriate to it.

The mismatch between school subjects studied (sciences) and psychological aptitude (divergent thinking) or intellectual ability are the factors which cause underperformance, lowered self-esteem and consequent lowering of aspirations. The fact that the studying of arts subjects might be a very important part of good training for higher level engineering work is rarely countenanced in Britain – because mathematics and physics are seen as (as far as engineering is concerned) obligatory – and in any case engineering tends to be viewed in the United Kingdom as a branch – the applied branch – of natural science. Design is arguably the central activity in engineering, the counterpart of research in the case of science. Yet design and engineering education have traditionally been to some degree separate in Britain, with many types of design taught in art colleges and attracting students with mainly arts qualifications. Post-Finniston engineering degrees are however increasingly emphasizing design's importance in engineering. Further, art college design education in the mid-1980s took some serious and arguably very long overdue steps in favour of industrial requirements. Even so these steps, however significant in a political sense, only represent a faltering start in a process of reintegrating engineering and art (Hudson, 1966, 1968, 1978).

The third explanation examines the relationship between social background and the choice of engineering career. A number of studies have repeatedly shown that engineering is a route to higher things for bright working-class boys. The lower aspirations may therefore be explained in terms of patterns of working-class or lower-middle-class socialization. Growing up in the working class means that bright children are at least partially isolated from the dominant social values of British culture and can thus more easily regard engineering as a worthwhile career. This may also account for the phenomenon that once at university or college, and away from the influence of the family, working-class students may change their minds about what constitutes a worthwhile career and alter direction toward research rather than careers in industry (cf. Cotgrove and Box, 1970).

Conclusion

Given the nature of the British tradition in education and the structure of engineering careers the social–psychological consequences for individuals who become engineers deserve some comment. Both sociologists and psychologists have contributed to an understanding of the general processes whereby people choose certain occupations and then learn the appropriate skills and ways of behaving associated with the occupation. These are known respectively as the processes of occupational choice and occupational socialization (see Ashton, 1973; Ashton, 1974; and Ashton and Field, 1976). Engineering is a particularly interesting case, since as we have demonstrated in this chapter, not only the general culture but also the education system and career opportunities conspire to make it a positive choice career only for a few people.

When someone chooses an occupation or drifts into a job, two processes are at work – individual and social. At the individual level the person has to consider a series of options or alternatives. Some of those options may suit his or her particular aptitudes, temperament, abilities, knowledge of the market, qualifications and so on, others may not. At the social level the basic values of society as well as the structural constraints of the availability of jobs, the demand for labour, the recruitment policies of firms and businesses and the nature of the labour market are important. Thus when someone enters an occupation this is the culmination of a process whereby the hopes and desires of an individual (the individual level) have to come to terms with the realities of the occupational market situation (the social level). The final outcome, the choice, stems from the complicated interplay of the individual and socioeconomic factors. Choosing an occupation involves the person reconciling the desired with the possible.

As far as occupational choice in engineering is concerned, the individual and social factors can be precisely formulated. At the individual level we have identified two sets of factors – those which repel individuals from engineering (school, education, higher education and employer attitudes) and those which attract (attitudes, aptitudes and social background). At the societal level we have identified a number of values – especially negative ones – as being crucial in respect of engineering. What the interplay between these involves is a major reconciliation process for the candidate for whom engineering is not the first choice. All of his education and training will involve a process of 'cooling-out' so that he comes to terms with the lower status position he has achieved (Goffman, 1952). For the first-choice candidate the process is not one of cooling-out so much as of coming to terms with the fact that it is unlikely that engineering is an end in itself – rather the means to the end of a career in 'management'. Either way, after the choice is made, much of the subsequent socialization involves coming to terms with the fact that society often mistakenly defines engineers as not good enough to be able to do much else.

Occupational socialization involves a number of distinct stages through which individuals must pass. The first stage is prior to entry, sometimes called the anticipation stage – which is where the process of occupational choice is made. Here the individual has certain expectations about the intended new job. These expectations are not pure fantasy but are based on information, both formal and informal, which the individual has gleaned from various sources. The knowledge of the new job will however be imperfect. This is the state in which the intended entrant to an engineering course or to an engineering career in industry finds himself.

The next stage in the process is referred to as the recruit stage. Here the individual is new to the job and the job is new to him. This stage is often accompanied by feelings of self-consciousness and doubt. This quickly merges into the next stage called reality shock. It is at this stage that the individual must align his imperfect anticipated knowledge with the reality of the occupational situation. This is the stage at which the largest number of those who are going to leave will do so. The majority survive however because in the process of occupational choice they have already undergone a process of aligning individual expectations with structural constraints. Here reality shock might be more accurately referred to as fine tuning of expectations. For engineers the fine tuning clearly involves the process we refer to as cooling-out – as individuals come to terms not only with the reality of engineering, but with the reality of the low status of engineering.

There follows the process of reorientation or of learning, whereby individuals develop strategies to help them cope with the situation (Elkin and Handel, 1972). The fully socialized member of an occupation is one who can no longer distinguish between strategies for coping and reality. The strategies become ways of living and working which eventually are regarded as unexceptional. As far as engineering is concerned two strategies for coping are noticeable. The first is to use engineering as a springboard into general management. For a fair number of engineers this is the solution which apparently proves most amenable. This we discuss in Chapter 9. The second strategy is a collective one whereby engineers band together in organizations, particularly professional bodies, and attempt to raise their status by imitating the professions of law and medicine. This strategy will be the subject of Chapter 10.

7 Theories of Work Motivation and Job Satisfaction

Introduction

In the previous chapter we argued that as a consequence of a particular value system rooted deeply in British culture there were certain social–psychological consequences for persons who became engineers. An area that has been particularly investigated in respect of the social–psychological effects of work has been that of motivation and job satisfaction. In this chapter we will examine theories of motivation and job satisfaction partly – but not only – to help suggest how these relate to the processes of occupational choice and occupational socialization described in the previous chapter.

Two main social–psychological factors were highlighted in the previous chapter as consequences of the educational and broader social arrangements which impinge on engineers. First, because engineers work in an environment which does not, in Britain, attract a great deal of prestige, they have to be socialized into this aspect of their work role. This involves at least some degree of reconciling hopes with reality. Second, we argued that the engineer has to be 'cooled-out' during his education and indeed in his work life in the sense that the low status of engineering not only has to be confronted in some way but its effects on the personality neutralized. These kinds of issue are explored in this chapter, with some emphasis on engineers as employees in the last section. However the main purpose of the chapter, apart from (hopefully) helping engineers to understand how they themselves may or may not be motivated, is to inform engineers working as managers, and to a lesser degree as colleagues, of others. These issues are important because they form the socio-psychological background to the study of motivation and job satisfaction. The theories of motivation and satisfaction only make sense in relation to engineering if the particular problems confronting many engineers are borne in mind.

Work, it must be stressed, is neither organized nor experienced in a vacuum. Many forces, amongst them historical, social, economic and political, have influenced its content, organization and rewards. Ways of designing jobs and occupations and of experiencing tasks and working relationships are remarkably varied.

We explore some of the major social scientific debates concerning motivation at work, focusing specifically on relationships between job satisfaction and effectiveness in the performance of tasks. In so doing we group relevant schools of thought into two major camps. One consists of those which suggest that managements should secure effective performance by being considerate towards employees. The other includes those which, conversely, suggest that effective performance should always have priority over employees' feelings, partly because they feel that the most fundamental cause of job satisfaction is effective performance. We have put the various schools of thought into two main camps in this rather arbitrary way for the sake of our present argument and we do not pretend to have done any of them full justice. Readers wishing to explore them in more detail are advised to consult the sources quoted in the text (see, for example, Vroom and Deci, 1970; Rose, 1975; Watson, 1980; Hill, 1981).

Job Satisfaction and Effective Performance – the Carrot or the Stick?

Can sociology offer something specific to those who want to motivate staff? The answer to the question is a provisional yes. When engineers and others are trying to get work out of staff, they are concerned with output and, at least in the longer term, employee satisfaction. These may be related for the sake of our argument by asking thus – to what extent does satisfaction with a job lead to good performance of tasks to be done and, conversely, to what extent does good performance itself constitute a source of satisfaction?

Answers will depend, in particular instances, on whether work is seen as a means to an end or as an end in itself. Unfortunately perhaps, job satisfaction cannot be observed, and people's reports of their feelings tend to vary suspiciously and consistently with job level. It took a little courage for Britain's Prince Charles to say a few years ago that he had a rotten job, and most people employed as lavatory attendants are unlikely to boast of how happy they are (Terkel, 1977). Other factors unrelated to job content affect our feelings. Gender is an example, as women have historically tended to report higher levels of job satisfaction than men. Performance is hard or impossible to measure with many jobs and occupations – think for example of the differences between the work of architects, surgical appliance fitters, design engineers, criminals, politicians, street traders and musicians. There is often a conflict between volume and quality of output in the long term and the short term, such as when profits come before investment and vice versa.

The many varied influences on performance include the characteristics of job holders (skills, attitudes, educational background and so on); resources available to them (finance, equipment, information); management and organization of work, and the environments of employing units (markets, attitudes of governments, legal systems and so on). Successful management demands at least some ability to control these influences.

Influences on job satisfaction are also varied. The following points, from numerous, mainly British and North American, surveys are indicative of this: married, older men are more concerned with job security than younger, single ones; married women are more interested in pay, as such, than single ones; expectations tend to rise with educational level; types of education and occupation influence expectations strongly, e.g. nurses learn to enjoy working with people, quantity surveyors to be problem-solvers, engineers to design and make things; smaller firms and departments are often happier places to work in than big ones; and attitudes to authority vary across frontiers, for example large Japanese firms tend to be paternalistic, Swedes and Germans respect high-quality products, Americans respect the ability to 'make a fast buck' and so on (Parker *et al.*, 1981; Hill, 1981; Child, 1984).

Factors normally associated with higher levels of satisfaction include seeing jobs through to their completion; control over the pace and methods of work; friendliness of co-workers, superiors and subordinates; autonomy and skill; permissive supervision; dealing with people; job security; and being consulted in advance about changes in ways of working. Older people tend to report higher levels of satisfaction than younger ones, and to be less concerned with the detailed content of jobs than with their social and physical environment. Two extremely important background factors are *variety* of tasks and *control* over work and its environment. The impact and content of work are more important to senior types of people, whereas subordinates talk more about pay and about work as 'something to do'.

These general findings must however be read against two other factors. First, there are marked cultural differences in attitudes to a whole range of things, not just work. Persons growing up in different cultures become socialized into the values of their culture. Their responses to questionnaire surveys concerned with attitudes to work will tend to reflect these broad values. Second, people's responses to questionnaires and their actual behaviour may be quite different things and simply because surveys have revealed the existence of particular attitudes does not, of course, mean the attitudes have a direct effect on behaviour. This is why, in general, those studies conducted by means of participant observation of work, i.e. where the investigator has actually done the job himself in order to acquaint himself with the values and norms of the work group, have generally been far more revealing than questionnaire surveys. A good, though now rather old, example of the participant observation method in studies of work is Roy's (1952, 1953, 1969).

The Twentieth-Century Schools of Thought: an Overview

The main schools of thought concerned with the performance of tasks, job satisfaction, attitudes to work and so on since the late nineteenth century may be depicted in Table 7.1. These will be described in detail in the next two sections (see also Mouzelis, 1967; Silverman, 1970; Watson, 1980; Hill, 1981; and especially Rose, 1975, on the first five schools, and especially Vroom and Deci, 1970, as well as Silverman, 1970, on the sixth).

It should be noted that scientific management was designed as an input to production, not as a part of social science. Its overriding concern was to make the management of production and operatives more 'scientific'. Rightly or wrongly it has long been associated with the fragmentation of tasks and jobs, with 'deskilling' and cost-cutting. From the nineteenth century onwards British public concern with employment conditions in mines, factories and other workplaces, suspicions about abuses of the techniques of scientific management, and problems experienced with overworked munitions workers in the First World War came together to produce a concern with the 'human factor' in employment and in particular to help stimulate interest amongst psychologists in physiological and psychological influences on work behaviour.

Such ideas influenced the American 'human relations movement' among social scientists and managers, beginning in the 1920s and arguably still working its details out today. Human relations opposed scientific management's view that people were primarily motivated by economic gain with one that men were motivated by affiliation or the need to belong.

Human relations writings effectively instituted industrial sociology as a discipline. The concerns of human relations and industrial

Table 7.1 Classification of Theories of Work Motivation and Job Satisfaction

School of Thought	Main Concerns	Period of Origin	Model of Man	General Context
1 SCIENTIFIC MANAGEMENT or TAYLORISM (after F. W. Taylor)	Efficiency, unsystematic management	1890s	Economic Man	Rapid growth of large-scale manufacturing in the USA
2 HUMAN RELATIONS (Elton Mayo and others)	Co-operation, social cohesion, efficiency	1920s	Social Man	The Depression and the 'mass politics' of the 1930s. Immigrants to the USA and their problems, particularly as first-generation factory workers
3 TECHNOLOGICAL IMPLICATIONS or SOCIOTECHNICAL SYSTEMS	Efficiency, social cohesion	Late 1940s, although mainly in the 1950s	Social Man	The increasing affluence of post-war Britain and the USA; also the increasing technical complexity of some types of work
4 NEO-HUMAN RELATIONS or ORGANIZATIONAL PSYCHOLOGY	Self-realization (of employees), social cohesion, efficiency	1950s	Self-Actualizing Man	As 3
5 the ACTION FRAME OF REFERENCE (as in the Luton studies)	Knowledge of British society (also a wish to show how men *act* and think *independently*; that they do not just conform to society's expectations)	1960s (some in the 1950s)	Complex Man – Sociological Version	As 3 and 4, plus the election of Conservative governments in the UK for the period 1951–64 ('Are we all middle-class now?')
6 EXPECTANCY THEORY	Knowledge of (individual) human behaviour	1960s	Complex Man – Psychological Version	Not relevant?

sociology have been quite diverse, but two major early developments can reasonably be isolated. 'Technological implications' researchers explored the varied technical and organizational contexts of worker and work-group behaviour. In doing so they helped to raise the sights of research from the small work group to the organizational level. Their work made a powerful contribution to the sociology of organizations and to the eventual establishment (around 1970) of a new discipline of organizational behaviour. Around the same time the second development, sometimes called neo-human relations, sometimes organizational psychology, was growing up. Here writers argued that, over and above their material and social 'needs', men also had a need for their work to provide them with opportunities for personal fulfilment, for what was called 'self-actualization'. They suggested that tasks and workplaces should be organized (or re-organized) to take this into account. Although there was much that was persuasive in this school of thought, it has been accused, with considerable justification, of preaching, of pro-management bias, and of various other forms of superficiality.

The two approaches which were mainly developed during the 1960s, the action frame of reference and expectancy theory, have a good deal in common. In particular, and in contrast to the scientific management, human relations, and technological implications approaches, neither is built up on assumptions about human needs being universally the same. In this sense both are methods more than theories. Both also strongly emphasize the independent sides of man's nature. Both have produced evidence which tends to contradict the rather static and rigid ideas about the nature of motivation which characterize the previous writings. Neither aims to change the world in any direct way. Both offer the possibility that a wider range of variables may be used to explain attitudes and behaviour at work than is the case with the other schools. The action frame of reference in sociology aims, among other things, to counter the tendency of the technological implications approach to define social phenomena at work as a set of reactions to systems of production, defined as a mix of hardware and methods of working and of organizing work. It also questions the assumption that all those who work broadly support the aims of those who employ and/or manage them, which had very often been implicit in the study of work organization and in scientific management and human relations (Goldthorpe, 1959, and 1966; Goldthorpe and Lockwood, 1963; Lockwood, 1966; Goldthorpe *et al.*, 1968, 1969 and 1970).

Expectancy theory in psychology has very similar concerns to the action frame of reference, although as a psychological theory concerned with individual motivation rather than a sociological one with much broader concerns, its narrower focus allows it to be tidier, more structured and more directly concerned with predicting behaviour.

At the start of the chapter we admitted our arbitrary way of distinguishing between these schools of thought. Not only can such classifications produce 'a narcissistic and abstract sociology of sociology', but they are inherently superficial simply because they implicitly compare and relate bodies of ideas and evidence which are often fundamentally different in their scope and implications (Hill, 1981). However our aim is not to produce a history of a broad body of thought and evidence, but simply to discuss it in a simple thematic review and in relation to practical concerns, with some historical understanding emerging as a by-product, We do need a theme for our ideas to have some significance, but we must again emphasize that the danger of a 'historical' approach is that it can so easily and often misleadingly suggest that each theory or school of thought is (a) at least partly a response to the perceived weaknesses of its predecessor, and (b) an advance on it.

The schools have had different aims, and although some have undoubtedly built on predecessors, the development of an understanding of the world of work has been very far from unilinear. Human beings have a natural tendency to regard the past as much simpler than the supposedly very complex present. People also tend to play down important differences between the varied settings in which past ideas were developed. Moreover some social scientists have hoped, a little naively, that their bodies of knowledge might be built up in the same sort of cumulative way in which knowledge has been built up in many fields in the natural sciences.

Most engineers exercise authority at some time in their careers. Using relevant material, we hope to help them realize that the dilemmas concerning the exercise of authority are not new and that engineers are not alone in being baffled by their complexity. Above all we hope that we show that while there can be no permanent or definitive answers in this area, indeed no 'answers' at all, there are several important questions which engineers can very usefully ask.

'Tender-Hearted' Approaches

Tender-hearted approaches broadly agree that 'a happy worker is an effective one', and that 'job satisfaction leads to good work performance'. These writers suggest that to motivate people one should first consider workers' needs, and second, and only then, the requirements of tasks to be done.

Elton Mayo, a failed Australian medical student who had switched to psychology and made his academic reputation working with shell-shocked soldiers after the First World War was the self-appointed father and publicist of the human relations tradition. This began to influence management teaching in the 1930s, and its effects are still seen in many general and personnel management and supervisory courses. It ostensibly prefers democratic management styles to autocratic ones. It grew up out of research conducted at the Hawthorne Works of the Western Electric Company in Chicago, which employed over 40,000 people in the 1920s and 1930s.

The 'Hawthorne Studies' lasted from late 1924 until the early 1940s. One major phase was the Relay Assembly Test Room study, begun in April 1927. Six female volunteer operatives were isolated and asked to work normally with an observer present. Their incentive payments and their hours and physical conditions of work were varied systematically, each with no direct or particular effect on output. Over two years there was however a steady rise in output: by June 1929 it was 30 per cent above the pre-experimental level. The researchers eventually concluded that the reasons must have been the increasing cohesiveness of the group, and especially the way in which the experiment itself made the subjects feel more wanted and important: subsequently this kind of experimental effect in social science has been known as the 'Hawthorne Effect'. The results of a large programme of increasingly open-ended interviews with employees which followed were interpreted as indicating a desire for more 'caring', 'democratic' management.

In November 1932 another major phase began. This was a study of fourteen male workers in a bank-wiring observation room who wired, soldered and inspected banks of telephone switchgear. It was found that the group was very clear as to what a fair day's output was. Almost all fourteen could exceed it, and they often did, but they then reported a false output figure close to the norm and worked more slowly on the next day. Persistent producers over the group output norm were derided as 'ratebusters'; those who produced less, falsely reporting that they had achieved the norm, were 'chisellers'. Supervisors colluded with the workers. Norms were met over a period of a week in any case. The researchers observed two cliques within the group, with the more productive one enjoying more status than the other. Workers were apparently spinning their work out because their jobs were threatened by the 1930s Depression, but the researchers' main concern was with how the *informal* organization of the group partly subverted the *formal* organization of their work by the management.

Human relations meant the 'discovery' of the worker as a social animal, desiring affiliation and approval as well as economic rewards, of the importance of informal organization and the less-than-rational character of motivation generally. The factory was depicted as a 'social system', not just a bureaucracy run along prescribed lines.

The tradition has been criticized for management-biased 'cow sociology', which sought to produce 'contented cows', looked after by caring but still fundamentally autocratic managements. The researchers have also been criticized for playing down their own evidence on the importance of economic rewards. The possibility that the workers in the bank-wiring observation room, for example, were restricting output as a means of preserving their jobs in the long run, rather than because of processes of group conformity, remains implicit in the research. The fact that many of the research subjects were first-generation factory workers and first- or second-generation immigrants to the USA and/or Chicago – people who might be expected to have an above-average desire to 'belong' – seems to have been discounted, at least in the more vulgar interpretations of the research. However the tradition has provided much ammunition for proponents of relatively enlightened management which makes itself aware of employee attitudes. Human relations ideas and assumptions underpinned the development of the industrial, occupational and social branches of psychology, represented the beginning of the sociological study of work, and fed almost directly into technological implications and organizational psychology.

The technological-implications approach (overlapping with, often called sociotechnical systems) focused on the interdependence of technical factors and social relationships and the experience of work. One major study is reported in Trist and Bamforth, 1951; Trist, *et al.*, 1963, dealt with the mechanization of coalmining in Durham after the Second World War and subsequent nationalization of the industry. New machinery was introduced in an apparent attempt to turn coal mines into mass production factories. Production fell to below pre-mechanization standards and absenteeism and labour turnover grew. Miners' skills and work-group autonomy and loyalties had been broken up by the new shift system. The geologically untidy, physically dangerous and unpredictable nature of the coalmines had been ignored: most factory assembly lines do not have to cope with flooding, gas, uneven ground and rock falls. Eventually the system was reorganized so as to restore many of the miners' old skills and autonomy, while continuing to use the new machinery, and both output and morale rose, the former to levels which were higher than ever.

Another well-known study which assumed that technical influences were central to understanding attitudes and behaviour at work was conducted by Blauner (1964) in the USA. He showed that 'alienation' was apparently low among workers with printing craft skills, slightly higher among machine minders in cotton textiles, was very marked among assembly-line workers in car factories, but was comparable to 'craft' levels in the technically sophisticated chemical-processing industry. For Blauner this suggested that alienation – although it is argued that most of what he was concerned with was job satisfaction or the lack of it – was due to the technical organization of work more than anything else. Moreover alienation seemed to be a function of the history of industrialization: craft skills represented the earliest type of manufacturing, textiles a subsequent one, car assembly an early twentieth-century development, while chemical continuous-process technology was conceived of as the most advanced manufacturing system at that time.

Horton (1964) criticized Blauner and others who tried to measure alienation in similar fashion on the basis of workers' subjective reports of their feelings for producing 'alienated accounts of alienation'. Like Silverman (1970) and Eldridge (1971), Horton criticized all of those who regard 'technology' (variously hardware, systems of production/operations, and/or ideas) as a – or the – major independent cause of attitudes and behaviour at work. For these critics, the technological-implications approach exaggerates the influence of technology compared with other factors such as employers' reasons for choosing systems of production and types of hardware and the varied expectations of employees concerning different work places and production systems. Like writers in the human relations *genre* they play down fundamental issues of power and conflict and of inequalities of opportunity.

Organizational psychology (or neo-human relations) was originally inspired by a theory of human motivation produced by A. H. Maslow (1943 and 1954). Maslow depicted 'human needs' as based on the principle that once basic needs such as food and shelter were satisfied, other needs emerge, such as safety, which in turn once satisfied lead to the emergence of further higher order needs. In this context Maslow's theory is usually referred to as the hierarchy of needs, and is conventionally represented as shown in Figure 7.1.

According to Maslow higher order needs cannot be satisfied until the lower order ones are met. So one cannot implement and express what

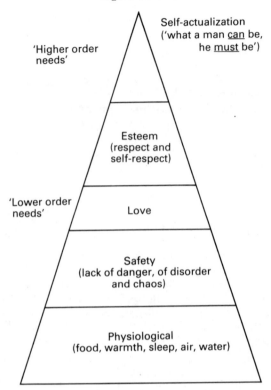

Figure 7.1   The Hierarchy of Needs

Source: Maslow (1943).

is truly unique about oneself, in work or play, or in one's personal life, until one is healthy, warm and safe, until one feels cared for, and until one feels and is respected.

This theory was applied to the study of motivation at work by Herzberg (Herzberg *et al.*, 1959; Herzberg, 1966; McGregor, 1960, and others). Herzberg advocated job enrichment and job enlargement, making jobs more responsible and complex, as did McGregor who also advocated participative management and decentralization of power. Herzberg questioned 200 accountants and engineers about events in their jobs which had resulted in a marked improvement or reduction in their job satisfaction. From the answers it was concluded that motivation-inducing factors *were different* from those which demotivated people:

'Satisfiers' or 'Motivators'	'Dissatisfiers' or 'Hygiene Factors'
Achievement	Company policy and administration
Recognition	Supervision
Work itself	Salary
Responsibility	Interpersonal relations
Advancement	Working conditions

The satisfiers relate to job *content*, the hygiene factors (preliminaries to 'healthy' motivation) to the *context* of work. The parallels with Maslow are obvious. The essence of Herzberg's approach is the idea that 'you cannot love an engineer into creativity . . . [it] will require a potentially creative task to do'. The point is that once an employee starts to think about such things as love, authority and motivation something must have gone wrong with the way the task appears to him or her. This is not to argue that social factors are irrelevant, rather it is a question of what makes people feel uneasy, or ill at ease.

McGregor wrote of Theory X (the stick) and Theory Y (the carrot). Theory X was the conventional management view of motivation: workers were usually passive or resistant to management, and they needed direction because they preferred to be led. It was suggested that theory Y was superior for motivating people whose physiological and safety needs were satisfied, and whose social, egoistic and self-fulfilment needs were prominent. According to McGregor's Theory Y workers had only become passive or resistant to change because managements had made them so. Workers had strong latent motivation and potential to be creative and responsible. Competent managements saw this and tried to arrange conditions and methods of work so that employees could achieve their own goals by directing their own efforts towards organizational objectives.

Both Herzberg and McGregor take the need for self-actualization as given and suggest that its satisfaction is stifled by autocratic and bureaucratic management. Their ideas are persuasive, popular with many managers and social-science-oriented teachers of management and influential although controversial among psychologists. There are five major criticisms. First, they tend to skate over issues of power and conflict: McGregor for example emphasizes the primacy of (existing) 'organizational needs' under Theory Y as well as Theory X. Secondly, it is assumed that human needs are universally identical, in all cultural settings and at all points in time. Yet 'wants and expectations are culturally determined *variables*, not psychological constants' (Goldthorpe *et al.*, 1968). Thirdly, as Maslow (1943) had himself pointed out, 'a satisfied need is not a motivator'. Thus once someone's physiological, safety, love and esteem needs are satisfied, it is not clear whether he will then endeavour to exercise and display creative impulses or to roll over and go to sleep! Fourthly, the theory is not good at explaining people like the painter, composer or poet starving in a garret, or soldiers and others who produce acts of great courage and imagination in conditions of great privation or danger. Their creativity occurs precisely when, or very often apparently because, their lower-order needs are unsatisfied (Alderfer, 1972). Finally, Herzberg's research methods were suspect in that when individuals are asked to account for their successes they naturally claim the credit, whereas if asked about bad events, they naturally blame others, for example 'the management'.

Scientific management and human relations were American in origin. Organizational psychology was not only American in origin but also its assumptions and values have a particularly American feel to them, more so than scientific management or human relations. Organizational psychology's model of 'self-actualizing man' resembles a 'sanitized' safely defused version of *homo faber*: it tries to push pseudo-Marxist ideas about motivation in a context which would traditionally be very hostile to them. Rush (1969) found that American managers' views of social science showed that Herzberg's and McGregor's theories had been the most influential in a country in which psychology and sociology are notably more established and respected than they are in Britain. In Britain, Watson (1980) has expressed distaste for 'behavioural science entrepreneurs' like Herzberg, who act as expensive consultants, but it should be remembered that the works of Maslow, McGregor and even Herzberg have been victims of their own past success. The popularizing and simplifying second-hand versions do little justice to the complexity of some of the original ideas.

The 'Tough-Minded' Approach

This is a much more mixed set of ideas than those in the previous section. Specifically scientific management, often associated with 'classical' management theories, the action frame of reference, and expectancy theory, are identified as tough-minded.

Scientific management was the brainchild of Frederick Winslow Taylor (1856–1915), the son of an upper-middle-class Philadelphian Quaker lawyer. Taylor began studying law himself, but had headaches and nervous trouble, and went into industry in 1874, taking up apprenticeships in pattern-making and machining. In 1878 he was offered an office job by the Midvale Steel Company, one of whose owners was a close friend of his family, but he went to work for it as a labourer instead. He became a machinist and was soon promoted to foreman. Having learnt all the workers' tricks of output restriction and 'soldiering' before his promotion, he succeeded for three years in increasing his men's output by using such conventional methods as sacking slackers, using blacklegs, and cutting piece rates.

Taylor was worried however about the harmful long-term effects of such methods. He felt that managements should try to discover the 'true' capacities of men and machines and then plan and control production 'objectively' in the light of such knowledge. Management should become a 'science' to sustain its authority. Wages should be both fair and seen to be fair. To develop and obtain recognition for such ideas, Taylor obtained an engineering degree by evening study, became a successful engineer, and later developed a coherent system for linking output and pay. He had a determined, uncompromising, even an obsessive personality. He certainly seems to have regarded toil as *the* paramount virtue and to have exhibited paternalistic and moralistic attitudes to a marked degree. It should also be noted that in his own mind at least Taylor believed that his system would improve wages and conditions for work people.

At Midvale Steel he developed and used his ideas and worked his way up into senior management. He was then invited to go to the much larger Bethlehem Steel Company and to help reorganize its management and methods. Bethlehem Steel was controlled by distant financiers rather than by locally based families, and it probably hired Taylor partly because of his engineering abilities and partly as a publicity stunt to make the company look go-ahead. He was appointed with nominally sweeping powers, but treated like an employee. In spite of some spectacular achievements he was sacked in 1901 with a one-sentence letter.

Between 1901 and his death in 1915 Taylor lived as a famous and controversial writer and consultant. He acquired important disciples like Gilbreth, who developed time and work and motion study and Gantt who tried to broaden the appeal of scientific management by taking serious notice of trade unions and the thoughts and feelings of workers. However other advocates of scientific management included hack consultants, who sought to increase output without raising wages, and who helped to get the idea a bad name.

Taylor seems to have assumed that men were economically rational, motivated by economic gain and rational when deciding how to pursue it; hedonistic, seeking pleasure and avoiding pain; and primarily self-interested. He saw labour as the main potential source of increased efficiency, with activity measurement as the main tool for improvement. His methods were to analyse all tasks in detail; to use time studies on their component parts; to analyse associated routines, such as stock control and office work; to select and train 'first-class men' to do 'a fair day's work' in 'the one best way'; to monitor results and to make any necessary adjustments. (A 'first-class' bricklayer, by the way, was not expected to have the same intellect as, say, a 'first-class' mining engineer).

On the one hand Taylor's work resulted in improvements to production, in piecework and other payments-by-results systems to get 'a fair day's work for a fair day's pay', a grand design for an industrial society with higher standards of living, and the scientific management movement. On the other Taylor had very simple, narrow ideas about relationships between pay, motivation and output, was not very appreciative of human weakness, and was unnecessarily negative towards trade unions. In his time in the USA the unions were mainly militant craft ones fearful of scientific management's potential impact on their members' skills. Yet Taylor was generally much more critical of remote, finance-oriented and arbitrary managements than he ever was about operatives, or even about militant unionism which he felt resulted from the former. His era was that of the 'robber barons', one of massive industrial growth and big organizations often manned largely by immigrants and run from afar by ruthless financiers. His ideal manager seemed to be in the John Wayne mould – tough but fair and human, and not just analytical. Several of Taylor's more influential followers came to accept the role of unions after advice from a United States government Commission on Industrial Relations set up in 1914 (Bendix, 1974; Rose, 1975).

Taylor's work and writings were produced between three and four generations ago. He was first and foremost an outstanding efficiency engineer, not an administrative, industrial relations or political theorist. He was undoubtedly an outstanding innovator and the first really powerful advocate of logic applied to production and to the systematic use of staff expertise in management. But he failed to appreciate the limitations of honest and technically enlightened but partly autocratic management. He was inadequately appreciative of variations in people's monetary and non-monetary values and their effects on motivation. Many of his techniques were often unreliable partly because of their internal difficulties and partly because of various forms of employee resistance. He had a rather simple faith in the potential of employer–employee co-operation. More fundamentally, he was naively conservative about the causes of economic conflict. He was guilty of a kind of 'scientism', in overestimating the possibilities of logic and reasoning in situations in which personal and group interests and values were predominant. He was narrowly 'technicist', in that he tried to impose his own particular philosophy of problem-solving in areas in which it could not cope adequately. Taylor did sometimes seem to want to treat people like machines – the 'first-class man' worked predictably, 'like clockwork', and harder the more he was fuelled with money. Most dangerously, his concern with logic and planning often meant that initiative was taken away from the 'workers', who knew most about particular jobs.

For all its shortcomings, Taylor's work has been extremely influential in many countries even among those who have tried to repudiate it. Motion study, following his idea of there being 'one best way' to do any job, and involving specialist study of the speed, comfort and so on of actions, has been developed in very considerable detail. So has time study, which sets standards of time and payment for particular tasks, and which together with motion study has been developed into work study and/or industrial engineering. Operational research, which adopts a broad detached view of tasks as a preliminary to attempting to make them more efficient, and which has increasingly used statistical and decision theories and computers, owes some important debts to scientific management. So too does ergonomics, which seeks to design machinery for efficient human operation. Like both operational research and industrial psychology, which studies the individual at work, ergonomics arose partly as a reaction to the crudities implicit in scientific management, as well as sharing the aim of improving efficiency. Personnel management has often followed Taylor's kind of philosophy in relation to the selection, testing, training and payment of employees, for example.

Taylor wanted 'system' to replace chaos and neglect in a context of economic expansion and the growth of large-scale organization. On this he was undoubtedly confused because dealing with chaos and lack of system is what in one sense provides the need for work. This philosophy would have been developed without him however: thus he was partly a willing focus and spokesman of a movement that would have come into being had he not lived. In its details his work and ideas could be both impressive and wasteful or useless. The philosophy has helped to create much wealth on the one hand, and to damage the psyches of some employees performing highly fragmented and repetitive tasks on the other. It has certainly created a liberal backlash among some social scientists who condemn the alleged alienating tendencies of scientific management. However he has often been blamed for acts and ideas for which he was not responsible. Taylor has been depicted as the high priest of the 'dehumanizing' assembly line which was only in its infancy in the last years of his life (Drucker, 1970). In any case far fewer people work on assembly lines in manufacturing than is often thought. For example Gallagher (1980) used official statistics to show that probably fewer than 5 per cent of Britain's labour force was so employed in the late 1970s: the proportion is probably smaller now.

The accusations of 'deskilling' often levelled at Taylor tend to ignore the fact that accountants and other financial specialists, rather than engineers, have been more responsible for the fragmentation of tasks in pursuit of greater efficiency (often without success). They also tend to neglect lessons learnt by developers of Taylor's ideas which are on balance more likely to enhance rather than destroy skills. Some social scientists have apparently forgotten how Taylor wanted to raise wages, to reform and democratize management, and to eliminate the stresses of work. According to Drucker (1970), Taylor offended intellectuals everywhere through being 'the first man in history who . . . studied work seriously', because he wanted to make dirty and exhausting manual tasks pleasant and psychologically as well as financially rewarding, by suggesting that a 'first–class' labourer deserved as much respect as a first-class professor, and above all by being successful. (Perhaps Taylor is best appreciated as the Elvis Presley of management theory! He was someone whom the self-appointed guardians of taste and the liberal conscience like to deride, but whose significance and impact cannot be overlooked. Of all the writers reviewed in this chapter he has had by far the most powerful direct effect on people's lives).

The 'classical' management theorists, writing from the late nineteenth century onwards, include the Frenchman Henri Fayol, the Americans Taylor, the Gilbreths (Frank and Lillian), Henry Gantt, James Mooney, Chester Barnard and Mary Parker Follett, and the Britons L. F. Urwick, E. F. L. Brech and Wilfred Brown. They have tended to emphasize the right to manage – to plan, organize, control, co-ordinate and motivate people – often a little more forcefully than Taylor did. Like Taylor they suggested, with various degrees of effectiveness, that the first duty of managements is to make employees deliver the goods and services, with their job satisfaction a secondary, albeit not unimportant, consideration.

The other two schools of thought which adopt a relatively hard-headed attitude to motivation are much more recent in origin and much less concerned with practical results than those just discussed. The first, the psychologists' expectancy (or path–goal) theory of motivation began to take its present shape in the 1950s: it was quite well represented in the writings of Victor Vroom (1964). It grew partly out of the organizational psychological ideas discussed above. However it emphatically refuses to accept that all or most people necessarily have the same or similar needs. It argues that our expectations about the results of our efforts largely determine what we do, and are hence the main influence on the degrees to which our jobs may satisfy us. Thus to pursue our goals, we weigh up the likely outcomes of the courses of action needed to achieve them in terms of their attractiveness and the probability that we can achieve them. So any situation has a number of 'action possibilities', each of which may lead to some form of reward. In considering a given situation, individuals evaluate the possible rewards and costs and choose the course of action which seems likely to lead to the highest valued reward. So for example the theory would predict that high job satisfaction will lead to low levels of labour turnover and absenteeism, because those concerned are motivated to go to work where important rewards are secured and/or important needs satisfied.

Expectancy theory suggests that it is sensible to think of performance leading to rewards, which in turn lead to satisfaction. Most studies comparing measures of performance and job satisfaction consistently show a low but positive association. According to Vroom, job satisfaction and performance are usually caused by different things. The former is closely affected by the extent to which jobs produce rewards, and performance is closely affected by the ways in which rewards are obtained. Job satisfaction arises from the provision of what people desire, insofar as they perform effectively to secure it. So if rewards cause satisfaction, and performance can produce rewards, then rewards form the key link between performance and satisfaction. In many ways the idea that performance causes satisfaction seems much more plausible than the reverse. For those in charge of others, it thus becomes more important to consider which people with what kinds of goal are likely to be satisfied by the jobs available, rather than to think first about ways of maximizing satisfaction. Neither is likely to be an easy thing to do in practice. For example pay is not an effective 'carrot' in many jobs, because it is so often obtained anyway, as a result of only moderate effort, and many supervisors and bosses are ignorant of the detailed requirements of tasks. Nevertheless the former probably remains the most sensible general philosophy to adopt.

The list of factors affecting performance and satisfaction allowed for by the theory is virtually endless. However they can be broadly classified in that intrinsic and extrinsic (to the job) sources of satisfaction are often distinguished by expectancy theorists. Influences regarded as important may include 'significant others' such as close family, friends and colleagues: someone could be offered a very good job in (say) South America, but might turn it down because his spouse did not want to move there.

The theory has been used to predict such things as people's choices of jobs and occupations with a degree of accuracy unusual in the social sciences. It is very useful for isolating specific influences, such as interesting work, good pay and career opportunities, which motivate people, or which demotivate them, such as repetitive work and unpleasant or dangerous working conditions, and for measurements of their relative importance. Expectancy theory is important here because it suggests that it is what we choose to do, and how effectively we choose to do it, that is the main influence on how 'satisfied' we will be. In use, it emphasizes the complexity of influences on satisfaction, performance, motivation and so on, depicting people as independent beings who actively pursue their own goals irrespective of managerial attempts to 'motivate' them according to the sometimes rigid tenets of Taylorism, human relations and neo-human relations.

The applications of the very fruitful 'action frame of reference' is best represented in the Luton or Affluent Worker Studies (Goldthorpe *et al.*, 1968, 1969, 1970). These examined the behaviour and attitudes to work, family life, education, managements, trade unions, politics and society in general of three types of employee living in Luton in the early/mid-1960s. Luton was chosen for the research because of its location in the relatively affluent English South Midlands: its workers would have, it was thought, rather different attitudes to work and to life in general than those in more settled and traditional areas. The subjects of the research were Vauxhall workers making cars on assembly lines; Skefco workers (nowadays SKF), making machinery and ball bearings, mainly using unit and batch production methods; and Laporte Chemicals workers, making chemical products, using flow process production. Some office workers were also studied. However most discussions of the findings have centred on car workers.

The aims of this sociological project were to explain how 'affluent', relatively well-paid manual workers, perhaps a 'new working class', fitted into society. Were they, by being able to afford mortgages, cars, washing machines and so on, becoming middle class and thus fulfilling the predictions of the so-called embourgeoisement thesis? Did they have the traditional 'them and us' image of society – or did they feel that the aims of managements were the same as theirs, or what? What did they want out of their jobs, their trade unions, political parties? What was the place of work in their lives?

More or less from the start the researchers argued that the human relations tradition had exaggerated psychological influences on attitudes and behaviour, neglecting wider social and political ones. They felt that the organizational psychology theorists had used too static a model of psychological needs, which in fact varied considerably according to differences in economic and sociopolitical environments. Also, those who had suggested that 'technological' influences on attitudes and behaviour were often crucial were also exaggerating and were guilty of a one-sided naive 'technological determinism'. The authors emphasized the desirability of considering all factors that might be relevant, especially, but not exclusively, the 'prior (to being at work) experiences' or 'orientations to work' of their research subjects.

Most of those interviewed did not express 'traditional' types of support for trade unions and the Labour Party. Instead of talking about working-class solidarity and fighting the bosses, they had emphatically instrumental attitudes, seeing the unions and the Labour Party mainly as means to the ends of better pay and conditions. The Labour Party was 'the best for people like us', not – as often in the past – a focus for idealism and solidarity. They tended to be privatized and home-centred in that their efforts at work were designed to benefit their families and they did not expect, or even seem to want, their jobs to be personally fulfilling. They were largely interested in extrinsic rewards such as money and holidays, not in such intrinsic ones as technical challenges or achievements.

This last finding served to criticize the idea of Herzberg, that once employees feel that they are reasonably well paid, once they are not dissatisfied with the context of a job, they will look to its content for positive satisfaction and creative fulfilment. Also, and in contradiction to the ideas of Elton Mayo and other human relations thinkers, most of the research subjects simply did not care whether supervisors were friendly or not. The ideal supervisor was someone who left them alone, and they often felt sorry for supervisors whom they felt had extra responsibilities without getting enough in the way of extra rewards. Against Blauner and others who had portrayed the assembly line as offering the least satisfying, most 'alienating' type of work, the researchers found their subjects to be reasonably satisfied. Also, levels of job satisfaction were similar irrespective of whether subjects worked for Vauxhall, Skefco or Laporte, in other words irrespective of differences in 'technology'.

The choice of Luton for the research and the 'self-selected' character of the subjects were very important. The subjects were often skilled men who had chosen to leave moderately paid but relatively satisfying types of work in poorer parts of Britain to do often much less satisfying but better paid jobs in affluent Luton. They were more mobile, in an era of full employment, than manual workers usually had been in the past. By choosing less well-regarded types of work than those which they had been trained for, they had lost a traditional kind of status but often been able to afford cars and their own houses. Thus they had often moved from skilled jobs or even from unemployment in the West Country or Scotland (for example), and from small, friendly firms which had not, however, offered good wages, or no wages at all.

The research showed that instrumental attitudes to work, employers and unions can override the impact of management/supervisory styles, technical factors, membership of work groups and trade unions, and the desire to obtain creative fulfilment at work. It demonstrated how it is usually vital to look outside, as well as at, the work place in order to understand the place of work in people's lives. The action frame of reference is, as suggested earlier, more of a general method and perspective than the sort of theory which offers a particular kind of explanation of something. It explains how people act in accordance with their experiences, and their perceptions of their situations, and that they do not simply 'behave' or perform particular roles scripted by others. Above all, its use has emphasized how men act to shape their lives, that they do not just behave in response to external constraints and forces: in doing so it has highlighted the rather passive, static assumptions about human nature which had been adopted in human relations, technological implications, and organizational psychological approaches. Its sociological version of 'Complex Man' has effectively complemented the psychological version of the same in expectancy theory.

Motivation and Job Satisfaction and Engineering

In their different ways each of the theories we have discussed can offer some insight into the issues of work behaviour, motivation and job satisfaction. However we will argue in favour of the action frame of reference as the most appropriate means of understanding.

Human relations and its derivatives offer a model of social man in which the need to belong, the need to be part of a group, is emphasized as crucial in understanding worker behaviour. Notwithstanding the objections to this approach to work which we highlighted above, the importance of the nature of social relationships and social rewards, indeed the social nature of work itself, remains an important idea. Work is a form of human behaviour and by definition human behaviour is social. This is a simple axiom, but one which highlights nevertheless an important quality of everyone's working life. Where human relations spotlights a particular issue of interest to engineers is in its identification of bad or poor human relations as a malady which can have ill effects on satisfaction or motivation. As we argue in our next two chapters and in Chapter 11 engineers are involved in a very complex set of human relationships which are, at least in Britain, relationships which pose problems for them. Not infrequently the relationships are conflictual and not harmonious and are the bases of power struggles rather than affiliation or gregariousness. By alerting us to the importance of 'in-group' loyalties and their significance in human conflict human relations seems very pertinent (even though the early human relationists could not see this very obvious element in their own reported findings). Engineers, like everybody else, will draw support from in-group membership and use this as a defence from others within the work situation.

But human relations offers another insight, this time into aspects of the social–psychological processes which impinge on employees, including engineers as employees. We have argued and demonstrated in this book that engineers do not receive particularly high status or rewards for what they do in the British context, and in our previous chapter we have suggested, by implication at least, that this is a problem which individual engineers have themselves to confront. However this may be a wrong implication to draw from the argument and it is here that human relations is informative. Human relations pinpointed non-monetary rewards as a significant motivator. Surely one of the rewards of engineering careers lies in the nature of the engineering task itself. In other words the activities of using skills and knowledge to make things and do things is in and of itself rewarding. The task is the reward, rather than any status or prestige accruing to the task. We marshall two arguments to support this idea. First the data reported in the previous chapter suggest that one of the principal attractors to engineering as a career was a previous interest in mechanical or technical things. Second, it accords with our model of *homo faber* which we argued was basic to our understanding of human social behaviour. There is a third argument which we draw out in Chapter 10 which shows that engineers have on the whole been luke-warm towards professional associations or even to broadly-based peer approval from other engineers in their work and have shown a commitment to the firm or organization that employs them. We believe that these three things taken together offer a significant explanation of why engineering is not a hotbed of dissatisfaction – the pleasure of the task often transcends other considerations.

By the same token the technological implications approach with its concentration on the technical and social arrangements in work situations offers a similar insight. Although society at large, or British society at any rate, looks down its nose at the engineering dimension, the socio-technical arrangements present an exciting and demanding challenge, strong in group loyalties and a concomitant disdain for outgroup members and their opinions, make engineering self-contained. The worries of academic engineers or pundits like us who are on the margins of engineering itself do not seem so often to form a preoccupation of the engineers themselves. From the organizational psychology perspective the engineering task offers the opportunity for self-actualization in a way that non-engineers do not appreciate. For those persons in engineering who do not find the task challenging there is usually a route out of engineering proper into management of some kind. The data reported in Chapter 9 offer broad support for this position in terms of the relatively high levels of satisfaction with their tasks and factors intrinsic to them reported by engineers.

Scientific management was the earliest systematic attempt at industrial engineering but strangely many of its practices seem to us on the whole to be anti-engineering. The attempts by Taylor and his followers to define the task very precisely and to apply what amount to crude auditing criteria to creative activity negate *homo faber*. Moreover Taylorism seems to put too much faith in analysis *per se*. Analysis of the work task may easily slip into a system of bureaucratic control not a system of creative manufacture. Of course, large-scale systems of production require both analysis and control, but it is all too easy for analysis and control to supersede the arts of making things or doing things, where analysis and control come to be seen as ends in their own right rather than means to achieve particular ends. We return to this question in our discussion of the production function in Chapter 11.

In the final analysis we find the action frame of reference and expectancy theory to be the most appropriate theoretical tools for understanding the experience of work of engineers and others. These ideas stress the notion that man is a thinking and acting animal who exercises choices and who is not pre-programmed or determined by crude 'mono' variables (money, social rewards and so on). People socially construct the world they inhabit. In other words people engage in symbolic activity; they think, they talk. The way that the world is seen and interpreted is mediated through language. People experience life as a series of biographical episodes, some of which take place in work organizations; others take place at home, at play, in love and so on. All individuals constantly interpret and reinterpret what is happening to them, what has happened to them, and what they imagine is going to happen to them. In this process of imaginative thinking they use language. Past events are accorded with present ones, new explanations for old life events are devised. People live in symbolic worlds and it is the meanings that they give to their world which are important to them, not the meanings of others that they do not know about or do not care about. The action frame of reference in industrial sociology recognizes this explicitly. For engineers and those who work with them it is not what others say or do about engineering which is significant at the level of experience, but rather the meanings that they as individuals give to their own experiences. These meanings are derived from socialization, and are derived from patterns of living at work and leisure. In Chapter 9 we open this issue up and examine in detail what it means to be an engineer. (For an elaboration of the theoretical ideas which inform the action frame of reference see Rose, 1962; Blumer, 1962; Berger, 1964; Mead, 1932; Rock, 1979; Field, 1974.)

8 *The Colleagues of the Engineer*

Introduction

The work roles of most engineers cannot be appreciated fully without an understanding of what their colleagues, superiors and subordinates do. In this chapter the tasks and roles of such people are discussed. Specifically we examine the work of marketing and sales, finance, personnel, construction, management services and senior managements. Then we offer a brief discussion of some features of the nature and daily exercise of power at work.

Engineering and Marketing and Sales

The commercial or marketing (or marketing/sales) function, if defined broadly, employs a very large number of people indeed, and the vast majority are not engineers. Marketing is the strategic activity of identifying consumers' needs and of working out how to satisfy them at a profit. Sales comprises the tactics of the operation. Marketing involves defining and developing products, studying markets (marketing research), and organizing selling efforts in the broadest sense so that advertising and pricing are planned for example. Sales management involves deploying a sales force in order to achieve sales targets agreed with colleagues in marketing, who receive feedback about customers and contacts by the sales representatives.

The American 'marketing concept', which has increasingly influenced management in Britain and other countries, aims to anticipate and satisfy consumer wants, rather than the older philosophy of 'we make it, you buy it' and 'if a man makes a better mousetrap, the world will beat a path to his door'. The marketing concept includes the idea of the 'marketing mix', most typically of the 'four Ps' of product, price, place and promotion. *Product* means such things as design, quality, appearance, and usefulness. *Price* is not just the basic price but includes related purchases, for example of road tax and number plates for a car, fittings for curtains, or missiles and spares for fighter aircraft. *Place* means the distribution channel, the ways in which goods and services are made available to consumers, rather than just location. *Promotion* involves advertising and publicity as well as personal selling.

Selling does not only involve consumer or technical products. Services are also sold. Advertising's aims are to attract, inform and persuade. Advertisements are used to establish new products, or to engage in brand competition or to maintain a product's image. Public relations is a very specialized activity in this field. It is usually an important function in large organizations in central and local government, the nationalized industries and public corporations, private industry and financial institutions.

Rosemary Stewart (1967) described many sales managers and representatives as 'emissaries' who spent a lot of their time with contacts external to their firms, who worked quite long hours, who travelled a lot, and whose working patterns were fragmented. Child and Ellis (1973) described marketing and sales as 'initiating' kinds of work, demanding a relatively open-minded and radical attitude towards established ways of working. Many studies of sales managers emphasize their very wide range of horizontal contacts inside and outside their units. Mintzberg (1973) contrasted the 'people orientation' of sales executives with the 'production orientation' of production managers. The former wanted products to be special, which was often inconvenient for the output figures of the latter. In general it seems that the relatively autonomous, 'innovatory' and sometimes free-wheeling/wheeler-dealing character of marketing and selling generates fairly consistent and often creative kinds of tension in sales and marketing people and in colleagues with whom they come into contact. Compared with production and finance, relationships of marketing/sales staff seem to be less hierarchical and more concerned with persuasion as an end in itself.

For some products and markets, personal selling is very important; but at the other extreme a great need for the product, or very high quality, or very low prices, or very persuasive advertising can turn selling into merely taking orders. Personal selling is usually more important when well-informed industrial customers are being dealt with than when customers are less informed members of the general public. There is a ready market for many routine services, for example for vehicle servicing, shoe repairing or hairdressing. More complex services such as those provided by management consultants, computer bureaux or insurance companies need a lot of detailed exposition if they are to sell. Some complex and long-lasting goods are almost sold as if they were services: examples might include houses and ships. Some selling consists mainly of getting repeat orders; other selling is mainly concerned with getting new business. Both types can involve 'milking the customer base', whereby salesmen persuade customers to buy more for use by more of their staff or departments, or to buy service contracts, or related products, and so on. All marketing and salesmanship has to take buyer power into account. Not surprisingly large buyers are normally more powerful than small ones because they are usually in a strong position for asking for discounts, special orders, additional after-sales servicing and so on. Salesmen also need to know about how far power is centralized in organizations that they are selling to: who is the 'MAN', the person with Money, Authority and the Need for the product? Marketing and sales staff are often in conflict with production. They often want to change products and delivery times to satisfy particular types of customer. The viewpoints of both parties are understandable but that does not help in itself. There is also evidence to suggest that engineers responsible for technical development are too often unaware of customer wants and needs. Further, fewer British sales and marketing staff appear to have engineering backgrounds (and to speak foreign languages) than foreign counterparts (Cunningham and Turnbull, 1981).

Engineering and the Financial Functions

Accountants are by far the most dominant group of occupations as far as the financial aspects of management are concerned. There are several different kinds of accountants, 'chartered', certified, cost and management, and public-sector accountants being the main ones. (We have put inverted commas around the word chartered because all of the professional associations for these four main types of accountant have been given Royal Charters, and because the Institute(s) of Chartered Accountants of England and Wales (and Scotland and Ireland) are probably only described using the word chartered because their bodies were the first to be so distinguished.) 'Chartered' and certified accountants belong to different professional associations but both types can work in private practice, checking and reviewing the accounts and 'books' of companies and self-employed individuals for taxation and other purposes, and acting as business consultants. Many 'chartered' and certified accountants also work as salaried managerial employees in private industry and commerce. Public sector accountants also have their own professional association and they produce accounts, analyse internal transactions for and give financial advice to their (public sector) employers such as local authorities, public corporations and the National Health Service.

Most cost and management accountants work in private industry and commerce and they are concerned with the provision of information for internal management purposes.

Accountants have three major responsibilities. Financial accounting includes auditing which means recording and checking the financial position of a unit in order to provide legally required documents at fixed (usually annual) intervals. These consist of balance sheets, profit and loss accounts, statements of the sources and uses of funds, directors' reports and (although this is not legally required even if it has become quite common) the annual reports to employers concerning company performance and prospects. To provide these documents, systematic records need to be kept of sales, costs of materials and parts, debts, wages and salaries, other overheads and fixed assets, and calculations of their depreciation. Balance sheets summarize assets and liabilities at particular stated times. Assets include buildings, land, plant, machinery, stock, work in progress, cash in hand and debts owed to the company. Liabilities include money owed to suppliers and to the government (corporation tax) and others, retained earnings (profits kept for reinvestment) and share capital (funds from shareholders). Profit and loss accounts assume that sales less costs equal profits. Profits are split into corporation tax, interest on loans, retained earnings and dividends to shareholders. Financial accounting is concerned, too, with cash flow, with the funding of current work, and also with obtaining funds for future activities. There are also financial ratios which are concerned with the return on capital employed (profit divided by sales, and sales divided by capital employed) to compare present with past performance. Financial accounting's emphasis is *external*; it controls the relations of a business with the outside world, with two of its key audiences being the Inland Revenue and shareholders.

Cost and management accounting's emphasis is on information for *internal* purposes, producing figures designed to help managements evaluate their activities from month to month. Statements are more detailed as well as more frequent than those provided by financial accounting. Internal transactions are the major concern. Costs of every kind are analysed, as are all possible sources of profit (plants, works, divisions, areas, and products). Particular jobs are analysed with regard to costs and profit. So too are particular ways of doing things, or 'processes' in the case of manufacturing. The information generated is used to help plan expenditure, to produce budgets as estimates of the cost of doing various things. It is then used to help allocate resources to departments and to monitor and control their use. The process of determining the validity of departmental claims is usually very political. Information on costs is also used for all kinds of one-off (but often fundamental or at least important) decision. These include 'make or buy' decisions concerning components and decisions about prices as well as strategic ones about markets to cater for and products to make, about research and development, acquisitions, diversification, rationalization and so on.

Engineers and accountants often appear to have little in common and to dislike each other. Some engineers seem to think that 'water runs in the veins' of accountants, and accountants sometimes believe that engineers are self-indulgent spendthrifts. Recent changes in engineering education are designed to make engineers more financially knowledgeable. Similar changes are *beginning* to take place in the education and training of accountants. But it is a mistake for engineers to regard accountants as natural enemies: there is an element of 'shooting the messenger' when accountants are blamed for reporting financial results of more basic problems.

Engineers and Personnel Managers

Some organizations have personnel departments as such and some do not, but all need to ensure that at least some important parts of personnel management – the part of management concerned with the employment and control of people – are performed. Personnel specialists are not as numerous as accountants, and there are far fewer of them than engineers and sales/marketing people. However they have risen to positions of some influence in many organizations over the last half-century or so, and only accountants and marketing people appear to have been paid as well in the last couple of decades.

There have been many lists of the varied tasks of personnel management. In recent years their main professional body, the Institute of Personnel Management, has grouped them under three main headings. Employee resourcing refers to the processes of employing people, to their organization as employees, the design of jobs, how people are recruited and promoted, motivated, paid, appraised and treated in other ways, and to their health, safety and welfare, and finally to the maintenance of employee records. Employee development refers to training and to management or personnel development, aiming to provide the organization with the people and skills that it needs and will need in the future. Employee relations deals with employees collectively and individually (Gospel, 1973). It is concerned with industrial relations, with employee participation in decision-making, and with communication which aims to transmit information of relevance and interest to employees.

An internal survey of members of the Institute of Personnel Management (IPM) showed that their major tasks were the recruitment and selection of employees, employee relations, training and development, and the payment of staff, followed fairly closely by manpower planning (of ways of obtaining and employing people), health, safety and welfare, and dealing, in one way or another, with legal matters and official regulations (Baillie, 1974). Most personnel departments are quite small, and the function generally seems to be more established and influential in larger units and in the public sector than elsewhere. Like white-collar trade unions, personnel departments seem to flourish in large bureaucratic organizations where the distinction between capital and labour is blurred and where managements are relatively neutral or favourable in their attitudes to employee rights.

Personnel jobs combine various elements of all of the areas of work described above, and they are often parts of the work of managers in other functional areas as, for example, when an accounts department is responsible for all aspects of remuneration or when production deals with most industrial relations matters on its own. This is one reason why personnel's position is often rather ambiguous and sometimes marginal.

Ambiguity concerning the personnel function seems to have four major characteristics. First, there seem to be many gaps between the bold prescriptions of personnel management textbooks concerning 'objective', 'rational' and structured ways of dealing with employees, and a much untidier and more political reality. Second, there are the very varied and uncertain boundaries between personnel and other areas of management (noted in the previous paragraph). Third, there is a historic tension between the welfare of employees – personnel's beginnings make it seem as if it was a form of industrial social work performed by caring upper-middle-class spinsters – and a much more hard-headed emphasis on control, for example by reducing labour costs and increasing efficiency through discipline. Fourth and finally, there appears to be something of an inbuilt tendency in personnel work to generate unduly elaborate and bureaucratic ways of organizing and doing things.

Personnel specialists seem to be more influential and more highly paid relative to other specialists in Britain than in West Germany, for example. The relevant differences may tell us something about an 'English sickness' of troublesome industrial relations, in which rule-making and collective bargaining tend to be at the work place rather than at the industry level (Maitland, 1983; Loveridge, 1983), and an arms'-length attitude to detail and production, and bureaucratic overmanning of managerial-level jobs (Lawrence, 1982; Child, *et al.*, 1983), using managers to manage managers.

The relevance of the idea of ambiguity in personnel work is underscored by data on personnel specialists' backgrounds. Estimates vary, but it has often been said that perhaps only half of practising personnel specialists belong to the occupation's major qualifying body, the IPM. The vast majority of those who pass its examinations have begun their working lives doing something else apart from personnel work as, indeed, have most personnel specialists (Baillie, 1974; Watson, 1977; Gill *et al.*, 1978; Melrose-Woodman, 1978; Glover, 1979). The backgrounds of personnel specialists are thus very varied although there have been several rather disparate tendencies for the occupation to attract women, social science and business studies and related graduates, former members of the Colonial Services and HM Forces, former trade union officials and shop stewards, and former engineers and technicians (the latter often into the training subfunction).

Thus personnel management is made up of a ragbag occupation in terms of its members' backgrounds. However weak the function it thrived in the 1960s and early 1970s, when economic conditions were good, when a great deal of employment and industrial relations legislation was going onto the statute books, when educational provision and social aspirations were rising. In the decade from 1963 onwards membership of the IPM doubled: it has continued to grow since, although not as fast until very recently. Traditionally personnel work has tended towards the 'initiating' rather than the 'controlling' end of the spectrum (like marketing/ sales and unlike accounting), but the balance appears to have shifted since the present recession began to bite seriously in the late 1970s.

Whatever the balance, personnel work is always likely to contain a large 'housekeeping' element – maintaining records of employees, training, recruitment, welfare, safety, remuneration and so on. Also, and as Armstrong (1984a) suggests, the tendency in many places for personnel or human resource aspects of management to be played down in favour of more urgent, technical or straightforward matters is always likely to exist. Experience, common sense, and research all suggest that the most successful (that is, useful) personnel departments will be sensitive to the true needs of their employers, and of managers in more central functions like production, to the practical rather than to the theoretical or gimmicky, that they will be unobtrusively constructive, and aware of the importance of competent presentation of ideas and monitoring of their implementation.

Engineers and Colleagues in Construction

The construction industry forms an important sector of the economy in terms of both employment and output. In the first half of the 1980s it was responsible for about 5.5 per cent of Britain's gross national product and nearly 5 per cent of total employment. At least about 9 per cent of Britain's chartered engineers work in construction, and this figure probably does not include all the engineers who do construction and related work in non-construction sectors (Engineering Council, 1983).

The construction industry is as different from manufacturing and commerce as they are from each other. Relationships between producers and consumers are different (usually less impersonal). The occupations employed are different. Ways of organizing tasks are different. The industry consists of three main parts, namely civil engineering construction, building, and (electrical and mechanical) engineering construction. Civil engineering means large projects – bridges, roads, dams, harbours, airports and very large buildings including factories, offices and high-rise flats. Building means housing (except in very large buildings), and smaller shops, offices, factories, hospitals, schools and other places of work: its scale tends to be small. Engineering construction means such things as the building of process plants and power stations and railway electrification schemes.

Many non-construction organizations contain departments which are mainly responsible for the maintenance of their buildings and land. These are known as direct labour organizations (DLOs) and they form parts of manufacturing companies, hospitals, banks, universities and many other types of organization apart from the local and regional authorities with which they are identified traditionally (Langford, 1982).

Small-scale building work is typically done by self-employed tradesmen and small builders who together (for example) buy land and build groups of houses for sale on it. As firms become larger their activities tend to diversify: they may build offices and schools as well as houses. Most building projects of any size involve two of the industry's three main types of professional, namely architects and (quantity) surveyors, the third type being civil and other engineers. Architects are responsible for design and quantity surveyors are technically knowledgeable financial controllers who estimate and monitor building costs. A high proportion of architects and of surveyors of all kinds, and many civil and structural engineers, are employed (or partners) in independent professional practices. Civil, structural and other engineers and engineering contractors are rarely involved in the 'building' side of the industry: normally the problems and the pay-offs are respectively too simple and too small. Many kinds of engineer, including electrical and mechanical, and not just civil and structural engineers, are involved in engineering construction.

Architects are often criticized for designing mainly with aesthetic criteria in mind and for being unaware of the needs of builders, of maintenance, and of costs. Quantity surveyors are sometimes criticized for being too narrowly concerned with the latter. Architects in the USA, West Germany and elsewhere often seem to be much more technically and financially skilled than British ones, and their tasks and skills include many of those of British quantity surveyors. Most construction in Britain, of all kinds, sees very little of the architect, and certainly not in a direct managerial capacity.

Criticisms of the British ways of educating and training construction professionals, and of organizing projects, focus on the idea of 'the missing manager' (Hislop, 1971). This refers to the fact that no single occupation has emerged to manage all aspects and stages of construction, from start to finish. Building, civil engineering, which employs quantity surveyors and architects too, and to some extent engineering construction, thus very often suffer from a 'too many chiefs' situation, although manufacturing is hardly a stranger to it either (Sorge and Warner, 1986).

However the British construction industry does tend to be more complicated than several foreign ones because of the high degree of horizontal specialization between the various professions and the various trades. (For example, as noted above, the tasks of architects and quantity surveyors are both performed by architects in the USA). Various schemes are widely used with varying success to cope with the situation. Project management, design and build, management contracting, package deals, and 'all-in service' are some of their names: what they all have in common is an attempt to impose order from the start on potential chaos, usually by putting one type of specialist in charge from start to finish (cf. Martin, Frenkel and Glover, 1987).

Management Services Specialists

These form a particularly varied group, who are usually employed to provide information and ideas to top job holders. In manufacturing, for example, some work in the head offices of large or medium-sized units; most are members of a quite wide range of specialist departments. They sometimes include people who would belong to other departments in other organizations, but who happen to be employed separately. For example, market researchers can work separately from the main marketing/sales function, perhaps in corporate planning units. Manpower planners and specialists in the design of organization structures, engineers concerned with planning design, production, or projects, or accountants involved in company taxation and financial planning, all fall into the same category.

In general management services specialists tend to work for departments which are highly specialized, manned by highly qualified specialists from quite a wide range of backgrounds, and relatively free-floating in the sense that their tasks are mainly advisory, and often relatively new. They are sometimes reasonably described as 'ideas men' or (less often) 'hatchet men'. Their jobs tend to be high on autonomy and low on people-centredness. The following job titles indicate the variety of their backgrounds and roles: work study manager, industrial engineer, operational researcher, management scientist, corporate planner, systems analyst, lawyer, economist, investment analyst, computer programmer, quality control manager, organization development manager, staff development manager, market researcher, financial planner and manpower planner.

The diversity of tasks done by such people suggests that they are likely to manifest varied attitudes towards them. Such attitudes thus tend to vary along a scale between being 'radical/flexible' and 'conservative/rigid' in relation to changes to established procedures and challenges to established authority. Accountants and quality control people seem to be more rigid than the average, for example, whereas operational researchers and systems analysts tend to be more radical and flexible. However there is probably an overriding tendency for all management services types to tend towards the latter, because the essence of their role is usually to initiate and innovate in one way or another.

Computer people, and the conflicts sometimes associated with their work, have been discussed by several writers (for example Pettigrew, 1973, and Child, 1984). They are often located outside the main functional areas, yet required to intervene in or work to alter them. Some studies have pointed to the relative youth and high qualifications of many computer specialists, and to their strong discipline, as opposed to organization orientation and to their role as 'change agents', all of which tends to provoke hostility from middle-level people in the more established line functions.

Management services people usually work in ways designed to inform decision-making by top job holders. Most studies of decision-making emphasize how it is very often a lengthy and political process and they often suggest that 'decisions' often happen to top job holders. Many acts of so-called decision-making are inevitable in the circumstances and/or products of inertia, supplemented by managerial rationalizing after the event. In fact many decisions concerned with simple, single issues, for which plenty of relevant information was available, and which could – to an outside observer – be made in a straightforward 'rational' way, still seem to take a long time to make and to be involved with political machination (on all these points see, for example, Mintzberg *et al.*, 1976).

Such evidence on decision-making casts great doubt on the idea that the broadly formal, 'rationally' conceived and organized kinds of information provided by computer and other management services people is *used* in a fundamentally rational way. It also sheds doubt on the idea that the world is increasingly run by experts, that possession of knowledge confers power in the way in which ownership of property used to. We have to find out whose property the knowledge is, and who employs the experts, to understand how knowledge is used. Many tasks of management services people are of the 'one-off' type in any case, meaning that it is easy for top job holders to monitor what they do. Yet at the same time the often highly specialized character of the information being handled provides the experts with opportunities to package it in ways designed to serve their own ends.

It is also important to remember that the types of knowledge and information used by top job holders are normally varied in the extreme, so that inputs to decision-making from the work of management services people may sometimes be fairly unimportant. They may think that they know what top job holders ought to do, but they will not usually know all the facts.

The important developments in computing and in information processing in general with which management services people are often involved and which have been affecting many aspects of employment are often expected to help managements to integrate and control activities more effectively than in the past. This is, in its turn, generally expected to produce smaller and more cohesive management structures. However, the availability of rapid means of processing information in fact simply provides managements with a range of new options for control and integration and the choices involved are partly dependent on the nature of particular employment sectors and partly on the attitudes and whims of particular managers (Child, 1984). A closely related but more basic problem for management services people concerns their outputs and the reactions of others to them. Successful executive work does not seem to be predictable, 'scientific' or 'rational' and attempts to make it so seem to involve the production of meta-concepts about work (for example, 'decision-making') rather than the outputs which job holders are employed to bring about (cf. Fores and Glover, 1976a).

Therefore 'ideas men' are unlikely to be very useful if all that they do is to theorize or produce speculative forecasts about events, or if they define themselves as members of a breed of generalist 'problem solvers'. They are likely to be more useful if they focus on more specific and specialist issues.

Top Job Holders: Boards of Directors

The traditional pattern of recruitment to top jobs in Britain has been to have generalists such as arts graduates from the older universities at the top, helped at the same level and especially lower down by larger numbers of professional and other specialists. This pattern seems to have been fairly typical in the 1950s but since then the backgrounds of top job holders seem to have become much more varied. So arts graduates plus the engineers and accountants who came up by the hard way of part-time study have been supplemented by engineering and science graduates, by graduates in management and business subjects including economics, and by holders of a variety of similar post-graduate and professional qualifications (Leggatt, 1978).

The values of members of large company boards tend to oscillate between middle-class ones which emphasize achievement, competition, profits and the devil-take-the-hindmost, and traditional, aristocratic, benevolent–paternalistic ones which more strongly emphasize duties and responsibilities of directors towards employees, shareholders and other groups. They generally lack the experience of working as, with, or managing manual or lower white-collar people, or any form of regular sociable contact with them; most of their contacts outside work tend to be with other managerial and professional types. When they have risen through the managerial ranks without benefit of family wealth or connections they tend to argue that their lack of financial involvement in their companies gives them a more

detached and rational attitude towards their work than their more
privileged counterparts. The latter often argue, however, that they
have a long-term commitment to 'their' companies which the former
may not have, while being just as competent and dedicated in an
everyday sense.

Neither group seem to identify as strongly with their companies as
do entrepreneur owner–managers whose involvement tends to have a
more total and dominating character. All tend to justify their
positions by arguing that the general good is served by efficient profit-
conscious managers who impartially and professionally balance the
often competing interests of owners/shareholders, employees,
managers and the wider community. They appear to feel that society
consists of two major groups – their own competitive, achievement-
oriented middle class, and the collectivist union- and state-centred
working class – plus an upper class which is no longer as powerful or
even really a part of the mainstream of society as it had once been.
They want to see a society in which there is more co-operation and less
class conflict, and in which 'doers' and creators of wealth like themselves are respected and listened to more often (Fidler, 1981).

Aspects of the intuitive and political character of executive work
were portrayed by Spencer (1981) in a study of British company
directors in board meetings. Some directors often appeared to be
relatively powerless and ineffective. Others were there as 'status
loaders', to legitimize and decorate proceedings which were usually
quietly (or otherwise) dominated by powerful executive directors.
Votes were rarely taken at board meetings, whose decisions emerged
from an atmosphere of gentlemanly restraint and consensus. In this
way differences in power and status remained implicit and no one had
to stand up to be counted: thus opinions were canvassed from the
chair, from whence a summing up and decision would follow. It
should be stressed that these are British findings; elsewhere open con-
frontation is probably more common or less well concealed (cf.
Lawrence, 1986; Mant, 1977).

Top Job Holders: Senior Civil Servants

The British Civil Service was reformed in 1870, when competitive
examinations were introduced to replace nepotism or other supposedly subjective ways of selecting people. The type of examinations used have helped to ensure that a high proportion of the top ranks
are staffed with Oxbridge arts-educated people (Dale, 1941; Kelsall,
1955, 1974; Dodd, 1967; Chapman, 1968, 1970; Halsey and Crewe,

1969; Sherif, 1976). The Fulton Report (1968) into the Civil Service proposed means whereby more graduates from less privileged institutions and backgrounds could be recruited but Kelsall (1974) has suggested that the subsequent changes to recruitment and promotion procedures were largely cosmetic and that they could well have the effect of solidifying past practices under the guise of changing them (cf. Kelly, 1980).

The British Civil Service makes a distinction between specialists who provide some particular service (like research) and generalist administrators who manage the service using specialist advice as appropriate. Specialists tend to be given, and to remain for most or all of their careers, in relatively comfortable, often interesting, middle-level jobs within their specialism. Generalist administrators are moved from section to section, very often before they develop any specialist expertise, although they do of course develop expertise in presenting arguments and data to their ministers and for internal and external consumption. Dixon (1980) cites careless overpayments of employment subsidies in the 1970s, inadequate monitoring of the plans and accounts of the nationalized industries due to lack of financial expertise, and legislative drafting errors as examples of the waste associated with lack of expertise at the top and its under-use below that level. For Dixon, the character of the policy-making administrators, highly intelligent but inexpert people who make life-time careers in the Civil Service, was the root of the problem. They were moved around too often to become fully useful in their postings. They were innocent of and reluctant to enter the world outside Whitehall. Even when they were seconded for outside experience, the knowledge so gained was frequently ignored on their return to the Service. Civil Service contacts have also told us that the kind of experience obtained may be of limited value in any case. This is because the civil servant on secondment often enters a kind of 'corporate mini-Whitehall', mixing largely with other top job holders and seeing little of the sharp end of business or the factory floor (cf. Pettigrew, 1985). It is probably too early to assess the long-term effects of the most recent attempts at improved efficiency spearheaded by Lord Rayner, formerly of Marks and Spencer, and a favourite of Margaret Thatcher.

Nevertheless the British Civil Service has been characterized by and indeed praised for its generalist orientation. The idea underlying generalism is that a finely-tuned mind can apply intellectual skills to any practical problem and if technical or specialist advice were required this could be obtained from a separate and effectively subordinate group of in-house experts. This idea was nurtured by the traditions of British public schools and Oxbridge education and sustained by the rapid turnover of political bosses of the departments of state. It is in fact something of a misnomer to refer to the senior civil service administrators as generalists because they undoubtedly develop specialized skills and specialist knowledge but it is also true that the system has produced a good deal of inefficiency (Kelly, 1980, chapter 4). One specific effect of this is that engineers and engineering have tended to be cast in the role of subordinate specialists *even* in more 'technical' departments like the Ministry of Defence. Engineer civil servants, it must be noted, are generally employed in the 'scientific' part of the Civil Service; another instance of engineering's subordination to science in the British public mind.

The Nature and Exercise of Power at Work

The concept of power has proved to be a tricky one as far as sociologists, political scientists and management theorists are concerned. According to Walsh *et al.* (1981) control and domination are more useful concepts than power for the study of relationships at work. This was because the exercise of power is only one of the more obvious ways in which people are controlled and dominated to their advantage or detriment. Normally, power is only exercised openly when values and/or interests are clearly opposed and when groups are in a position to confront each other. Sometimes groups can get involved in conflict against their material interests, because someone has offended their values.

Yet it often is crude power which, in the final analysis, makes people do things. It is important to remember that power and authority are not the same. Having the right to make someone do something against their will is not always the same as being able to force them to and vice versa (Weber, 1947). Without power, authority is vulnerable, and power without authority is dangerous. The power and the authority that a given manager can use normally depends heavily on circumstances, on knowing what 'the system' is and how to work it. Every individual or group has resources of one kind or another. The ability to use them depends on their importance to the organization, their relative scarcity, the skill, determination and effort that goes into their use and the help given or resistance offered by others. Every manager should have some political awareness: neither the completely honest and loyal servant of the board nor the completely neutral expert is likely to prosper for long in most settings.

A well-known list of sources of power was produced by French and

Raven (1968). Reward power is the ability to give people what they want (the 'carrot'). Coercive power is the ability to thwart or punish them (the 'stick'). Legitimate power is conferred by subordinates on people whose position or attributes they accept and respect, people whom they feel they should obey. Referent power is based on subordinates identifying positively with a superior, because of what they represent. Expert power is based on relevant ability and knowledge. Armstrong (1984a) developed a very similar but longer list: access to people with power, control over information, control over results (over what is achieved or is able to be achieved), control over resources and over rewards and punishments, control over expertise and the ability to get others to identify with oneself. Mechanic (1962) argued that power consists of information which people want, people who can get things done, and instrumentalities, things which people want or need. In all these definitions, power is seen as consisting partly of material resources, partly of information and ideas, and as being mobilized with varying degrees of efficiency.

At work, and as in war and in sport and games, skilful mobilization of a relatively weak power base can often be more successful than incompetent mobilization of a strong one. The threat or the use of force to obtain desired actions is usually effective at first, but in the longer run it may be self-defeating due to the ensuing resentment. Persuasion is usually preferable, although it often needs the background support of possible coercion or of referent or legitimate power. A very common method is to mix force and persuasion by exchanging favours or services for desired actions. Managers with the power to sanction pay rises or promotions can get staff to work unpaid overtime more easily than if they did not have it. Managers often form temporary alliances with each other on an 'I'll scratch your back' basis. More formally, firms agree to obtain specified percentages of raw materials or parts from particular suppliers, provided that delivery schedules are met. In the longer run successful managers implement their 'agendas' by building up 'networks' of relationships with people whose acquiescence or support they need (Kotter, 1982a, 1982b). Thus they aim to reorganize their environments by changing the ways in which people think and behave. They avoid situations, for example, in which particular subordinates or small groups of them are given more power than others. They ally themselves, formally and informally, with people with similar interests or mental outlooks.

Earlier data obtained by Kotter showed that successful managers used their power openly and legitimately, mainly on the basis of consistently-displayed practical and leadership abilities. They were sensitive to the types of power which different types of people respect.

Experts, for example, respected expertise. They developed all their sources of power and did not rely too often on a particular type. They sought jobs and tasks which gave them the opportunity to acquire and use power, continually finding ways to invest the power they had to obtain an even greater amount. They enjoyed influencing others, but did so in a self-controlled, mature way without using power impulsively or for their own aggrandisement (Kotter, 1973). One might also add that they accept conflict as being an inevitable feature of progress and change, and that when it cannot be eliminated by joint development and implementation of solutions, either compromise or peaceful co-existence can be acceptable alternatives. Allen *et al.* (1979) found that politically effective managers were felt to be 'articulate' and 'sensitive' above all, and also socially adept, competent, popular, extrovert, self-confident, aggressive, ambitious and devious. Intelligence, logic and being a good 'organization man' were also felt to be useful, albeit not quite as much. Even if these lists of adjectives confuse the reader as much as they do us, they still give a feel for the nature of political effectiveness!

Many management textbooks argue that managers should plan, organize and control events using the formal hierarchy of power and authority, by developing and using leadership skills, and also by thinking of the organization as a predominantly co-operative system of interdependent sub-units and with more or less predictable relationships with other organizations. However all kinds of evidence on the reality of managers' jobs emphasize their extremely varied, often unpredictable, political and often reactive nature: much managerial work consisted at best of 'coping' with events rather than controlling them.

Managerial power is of course complex and varied – like managers' backgrounds and jobs – but it is real. Management 'works *with* labour, but also stands over it' (Storey, 1983). The fundamental role of management in business is control of labour in the name of efficiency with managements' ultimate aims including profit and the preservation of its privileges. As production and other techniques have become more sophisticated, organizations have grown and they have employed more and more highly qualified experts. But that may only mean that the powerful have become experts, rather than the experts powerful. 'Management' wields great power but individual managers often have very little. They may often not even know where to find that which they do have. According to Pym (1975) the growth in the scale and therefore in the complexity of management has made decision-making more of a collective than an individual matter. Objectives and ways of assessing individual performance have become harder to pin down; individuals have found achievement increasingly difficult even to define. The 'response to this crisis had been to deny it' by inventing a 'massive arsenal of . . . meetings, objective-setting games, appraisals and assessments, information systems . . . graduate recruitment programmes, management sciences and job redesigns which provide the delusions of purpose, tangibility, personal responsibility and performance'.

Conclusion

Engineers in the course of their daily work have to deal or may have to deal with a range of other different occupational groups some of which we have highlighted in this chapter. They will also have to work with technicians, craftsmen, skilled manual workers, semi-skilled and unskilled people as well as various clerical employees who provide services like typing, telephoning and messages within employing units. In this chapter we have concentrated on the groups with whom the engineer is likely to have a relationship of near equality or be subordinate to, rather than those whom the engineer controls or directs. There are two arguments to make concerning all these relationships. First the values, beliefs and pattern of behaviour within each of the specialisms we have discussed will have a character of their own which will reflect their own occupational culture as well as specific local circumstances. In sociological terms we refer to this as a sub-culture within a total culture. Sub-cultural groups based around occupations (and this applies equally well to manual and clerical workers as to top job holders) not only develop their own sets of values and norms but also their own ways of doing things, their own private jargon and 'habitual way' of thinking about problems. Sub-cultural groups in any society or total culture will not only come into conflict with other sub-cultural groupings from time to time but develop ideas about 'out-group' members. This is a feature of all social life and it is not therefore surprising to find it manifesting itself in employing units. Engineers have to operate in such a situation and given broader British values it might be expected that engineering would enjoy less power within most employing units than for example finance specialists (Ingham, 1984).

· The second point is that the web of relationships between the different groups is extremely complex but is not static. The division of labour, apart from being a system for getting things done, is also well understood as more than just a structure within which things happen; it is also a continually evolving process. It will change as internal and external conditions alter. These changes along with conflicts are potentially powerful influences on individuals' feelings about their lives and the meanings that they give to the particular circumstances in which they find themselves working. In the next chapter we explore some of these influences.

9 *The Organization and Experience of Engineering*

Introduction

This chapter is concerned with engineering work and with evidence of engineers' experience of it. Following a brief introductory statement about the relative size of major British centres of employment, the first main section goes on to describe several features of construction and manufacturing in Britain: their economic roles, size, employment patterns, performance and products. We also make a number of contextual points about the ownership, finance and development of British industry. The next section discusses public sector engineering and a number of government and EEC policies designed to support and to otherwise affect industry. In the third section we discuss some general features of the organization of engineering work and we describe the main public, but non-governmental, bodies which represent the interests of engineering employers. In the last section we describe the results of a number of studies of the subjective experience of work of engineers. Taken as a whole this chapter offers a picture of some major aspects of engineering, stressing its variety and its achievements as well as its weaknesses and its frustrations.

Construction and Manufacturing Sectors

Data on the construction industry show that it accounted for 5.4 per cent of Gross Domestic Product and 5 per cent of employment in 1984. About 54 per cent of this value consisted of new work, and the rest was in repairs and maintenance. About 43 per cent of all the work was in housing. Apart from the 5 per cent of the labour force employed in construction (about 1 million people) there were another 460,000 self-employed. Most construction firms are very small, although there were about 95,000 firms employing two or more people, 95 per cent of them employed less than 25.

Construction involves the design, fabrication, alteration, repair and maintenance of buildings, highways, drainage and sewerage systems, docks, harbours and canals, sea defence works, offshore structures, electrical wiring, heating and other installation work, and structural work connected with power stations and telecommunications. Much maintenance and repair work and some new building are undertaken by public authorities. These often employ or work with the help of civil and structural engineers, who also work for larger building and civil engineering firms and independently as consultants. Most design is by architects or consulting engineers.

Census data have suggested that about one in ten graduate engineers and a similar proportion of holders of engineering Higher National Certificates and Diplomas work in construction. The CEI and Engineering Council surveys have produced figures between 6 and 11 per cent but a separate category of consultancy has yielded figures almost as high. In fact both the Census and CEI and Engineering Council figures may underestimate the proportion of engineers in construction because much construction work as defined above is done in sectors outside construction – private building, civil engineering and engineering construction – as it is usually understood. Not only is construction work undertaken there and by local and central government, but also in the public utilities and corporations, the nationalized industries, petrochemicals, manufacturing, mining and so on. There is a further problem with CEI and Engineering Council figures in that surveys of their members automatically exclude non-affiliated engineers, a significant segment numerically of those engaged in engineering, and because engineers working in construction tend to be members of Engineering Council institutions more often than their counterparts in most other sectors.

Construction has been a fairly useful source of foreign earnings for Britain in recent years: fees earned by consulting engineers, architects and quantity surveyors have all helped. This has been partly due to the post-1973 increase in the incomes of oil producers (as in Saudi Arabia), and the recession at home, which forced construction firms to look overseas (Sargent, 1978). Employment in construction has suffered very badly in the recession; the sector is a traditional indicator of booms and slumps, tending to prosper exceptionally well in the fat years and to suffer inordinately in the lean ones.

The proportion of Britain's employed labour force in manufacturing is now about 26 per cent and some 65 per cent of the employed labour force is engaged in service sectors (HMSO, 1986). Manufacturing industry accounts for just under 25 per cent of Britain's Gross National Product. About two-thirds of the country's visible (physical) exports consist of manufactures and semi-manufactures. Since the early 1970s both exports and imports of manufactures have risen, the latter by more than the former. Exports rose most significantly in electrical and electronic engineering and in chemicals. Apart from iron and steel and shipbuilding most of Britain's manufacturing industry is in private hands and current (in 1987) government policy is to reduce the state ownership of industry. Over half of those employed in manufacturing work for concerns with 500 or more employees; 28 per cent work for firms with at least 1,500. In 1980 the 100 largest private companies employed 37 per cent of the country's employees in manufacturing. Metal and vehicle manufacturing and other kinds of engineering concern tend to be larger than those in most other kinds of manufacturing. Types of manufacturing in which a few very large firms account for most of the output include motor vehicles and their engines, tobacco, railway vehicles, asbestos, cement, lime and plaster, and man-made fibres. Big firms also tend to dominate shipbuilding, aerospace, spirit distilling, watches and clocks, and other kinds of vehicle apart from those just mentioned. Imperial Chemical Industries, Unilever, the General Electric Company, Ford, Austin Rover, and Thorn-EMI are among the best-known of the largest British manufacturing companies as measured by their turnover.

Table 9.1 shows the relative size of manufacturing sectors in 1983 and 1984.

After the rise in oil prices of 1973 manufacturing output fell sharply, but grew gradually again from 1975 to 1979. Output fell again between then and 1982. These fluctuations affected most of the industrial countries but until 1984 high interest rates and a rise in the exchange value of the pound affecting exporters made things particularly difficult for Britain. Output began to pick up in 1983 helped by a fall in the world price of oil. In 1982, a 'trough' year in the most recent cycle of investment in manufacturing, total direct investment in manufacturing was £5,183 million, consisting of £3,950 million on plant and machinery, £732 million on new building, and £501 million on vehicles. The sectors which invested most were chemicals and man-made fibres (£705 million), food (£670 million), electrical and instrument engineering (£520 million) and mechanical engineering (£489 million).

Table 9.2 shows the export and import performance and, in a little more detail than in Table 9.1, the relative sizes of the main manufacturing sectors.

Various facts about the different manufacturing sectors are of particular interest. It is important to remember, now, that although many

Table 9.1   *Manufacturing: Net Output, Index of Production, and Investment*

Sector	Net Output £ million, 1983	Index of Production (1980=100) 1983	1984	Gross Domestic Fixed Capital Formation £ million, 1984
Food, drink and tobacco	8,913	101.0	102.0	1,191
Mechanical engineering	8,048	87.4	87.1	} 898
Other metal goods	3,433	94.5	100.1	
Electrical and instrument engineering	8,334	108.1	123.4	873
Chemicals and man-made fibres	6,254	107.0	113.5	1,062
Motor vehicles and parts	3,423	83.9	81.3	} 855
Other transport equipment	3,586	95.0	91.1	
Metal manufacturing	2,386	104.4	109.9	334
Other minerals and mineral products	3,437	93.9	95.1	457
Textiles	1,873	91.3	93.8	} 322
Clothing, footwear and leather	2,105	97.4	101.5	
Paper, printing and publishing	5,910	92.1	95.9	} 1,076
All other manufacturing	4,812	95.2	98.9	
Totals	62,514	96.9	100.6	7,068

Source: United Kingdom National Accounts, 1985 edition.

are commercially sucessful and innovative, many are not and most are nothing like as important in world terms as they once were. Iron and steel making has been in relative and absolute decline since the 1960s. Since the 1970s, the British iron and steel industries have become far more efficient (and employ far fewer people) than was the case then. They now have an outstanding recent record of productivity by European standards. Even so the overall record is not outstanding by world standards; nor is output in relation to population other than mediocre for a mature industrial country. About 85 per cent of production by volume and about 70 per cent by value comes from the British Steel Corporation: private firms make the rest. About 30 per cent of the finished steel is exported directly; more is exported as part of other finished products. Major non-ferrous metals processed and fabricated in Britain include aluminium, copper, lead, zinc, nickel, tin and titanium.

The production of building materials such as bricks, tiles, chimney pots, plaster, slate and concrete is as sophisticated in Britain as it is

Table 9.2   *Manufacturing: Labour, Sales, Exports and Imports in 1984*

Sector	Labour (000s)[1]	Sales £m[2]	Exports £m	Imports £m
Metal manufacturing	208	7,887	3,766	3,947
Non-metallic mineral products	194	6,923	985	820
Chemicals	326	18,886	8,233	6,551
Man-made fibres	15	575	419	346
Mechanical engineering	772	17,516	7,389	5,424
Other metal goods	379	8,600	1,093	1,246
Office machinery and data processing equipment	73	2,609	2,787	3,838
Electrical and electronic engineering	648	9,336	3,964	5,710
Instrument engineering	109	2,148	1,112	1,384
Motor vehicles (including trailers and caravans)	157	6,076	1,720	4,579
Motor vehicle parts	133	3,316	1,836	1,478
Other transport equipment (including ships, railways, aerospace)	293	5,561	3,811	2,532
Food manufacturing	485	25,012	2,041	5,995
Drinks	104	4,903	1,244	913
Tobacco	22	1,854	409	87
Textiles	231	5,294	1,620	2,927
Footwear	50	878	146	656
Clothing, Hats and Gloves	194	3,341	659	1,032
Made-up textiles	24	685	90	173
Fur goods	4	35	145	98
Leather and leather goods	24	666	232	347
Timber and wooden furniture	202	4,878	356	2,185
Paper and paper products	143	6,198	741	2,849
Printing and publishing	339	8,612	725	489
Processing of rubber and plastics	173	6,280	1,380	1,654
Toys and sports goods	25	429	207	413
Other manufacturing	48	1,383	1,013	1,298

1   Employment figures are for June 1984.
2   Sales figures represent sales by manufacturers.
Source: Britain: An Official Handbook, HMSO, 1986.

virtually anywhere else. British pottery is often of a high standard. Britain's glass industry remains one of the world's largest.

The chemical industry is Europe's second largest. Imperial Chemical Industries (ICI) is the fourth largest chemicals company in the world, accounting for about 30 per cent of British chemicals production. Some British manufacturers of pharmaceuticals and man-made fibres have been in the forefront of innovation in their fields.

Mechanical engineering officially includes all kinds of non-electrical machinery, machine tools, mechanical handling equipment, construction equipment, industrial plant and industrial engines. It contains several industries, in fact, and there is a very good case for breaking the category up accordingly. Media sources which, in similar vein, refer to 'the' engineering industry are also culpable. More than half of the output of mechanical engineering is bought in Britain: heavy equipment is bought by steelmakers, manufacturers of chemicals, and the nationalized fuel industries among others. Most kinds of industrial plant and steelwork are made in Britain. So are most kinds of agricultural machinery and machine tools although manufacturers of the latter have had some fairly spectacular problems since the 1960s. Virtually all kinds of process machinery are made in and exported from Britain; they include machines for bottling, bottle-washing, canning, labelling and packing; machines for use in food, drink and tobacco production; all kinds of industrial chemical equipment; furnaces, ovens, kilns, and gas, water and waste treatment plant. British manufacturers are net exporters of these as well as of all kinds of plant used in construction, and of many other kinds of engine, tool and other devices and appliances in mechanical engineering.

Office machinery and computing equipment tend to be provinces of multinationals but some British manufacturers are in the forefront of development and growth. Electrical and electronic engineering is a fairly strong sector; one in which Britain tends to export basic equipment and capital goods but to import electronic components and consumer goods, and domestic electric appliances. British manufacturers include some of the world's more innovative ones: defence, aviation, shipping and health electronics are some of the main areas in which significant achievements have been made. The related instrument engineering sector is fairly strong.

The main motor car and commercial vehicle manufacturers are Austin Rover, Ford, Vauxhall and Talbot: they account for all but about 1 per cent of British-made (or -assembled) cars and 2 per cent of commercial vehicles. Only Austin Rover, however, is really a British company; Ford and Vauxhall belong to American multinationals and Talbot to a French one. Only Austin Rover, too, is more or less publicly owned: the Conservative government's intention is to return all or part of it to private ownership. These 'British' motor manufacturers have been fighting back against European and Japanese imports with some success in recent years. Motor-cycle manufacturing has however largely disappeared from Britain since the 1950s and 1960s;

the manufacture of pedal cycles has nevertheless been doing quite well recently.

British railway equipment-making is quite small in scale by international standards, but quite strong on exports. Aerospace is also fairly successful in Britain, partly because of the country's relatively high levels of defence spending, and one of the world's largest and most comprehensive aerospace sectors. Rolls-Royce is one of the world's top three aircraft engine manufacturers. British Aerospace was effectively returned to the private sector in 1981 after three years as a public corporation. Some of its products are among the world's more advanced aircraft. The Jaguar tactical strike aircraft, the Harrier vertical and short take-off and landing aircraft, the Hawk trainer, the Tornado multi-role combat aircraft, the European combat aircraft, Concorde and the A300 Airbus and its derivatives are all products or partly products of international collaboration in which Britain has played or will play a major part. Britain is also involved in the manufacture of short-range commercial and light aircraft and of avionics and other aviation equipment and in space systems.

Although British shipbuilding capacity is very much reduced indeed, both in absolute terms and (especially) in relation to foreign capacity, British shipyards are still able to make almost all kinds of vessel and offshore structure. Some of the methods used are very up to date by most foreign standards, using covered-berth 'ship factories' for example. The public sector (British Shipbuilders and Harland and Wolff in Belfast) produces about 97 per cent of Britain's small output of merchant shipping, most of the Royal Navy's ships, and all slow-speed moving diesel engines. In 1982 the value of British exports of ships and ship repair services was about three times that of the country's corresponding imports, and nearly a quarter of the merchant ships produced in Britain in 1982 were exported. Private sector firms tend to concentrate on the construction and repair of smaller vessels like tugs, coasters and fishing and leisure boats. British Shipbuilders and private firms do a lot of consultancy and submersible and semi-submersible and other types of construction work for the off-shore oil and gas industries.

Britain's food processing industry has grown partly because of growth in the demand for convenience foods, and partly because its work has helped to increase domestic production of food from under 60 per cent of home consumption in 1970 to about 80 per cent now, and to increase Britain's food exports. The value of Britain's exports of alcoholic beverages was nearly double that of corresponding imports in 1982. Over 80 per cent of the whisky produced – largely in Scotland – was exported, especially to the USA, Europe and Japan. British production of gin, vodka and soft drinks has increased markedly in the last generation. Among beers the demand for lager has grown significantly. Almost all the tobacco products sold in Britain are manufactured in Britain, and Africa, the Middle East and parts of Europe and South America provide important markets for British cigarettes.

British textile manufacturing, as elsewhere in Europe, has been severely affected by imports from the developing countries. Cotton has been greatly affected in the same way but at least wool exports in 1982 were nearly two and a half times the value of imports. Increasing use of man-made fibres made by chemicals manufacturers has blurred the boundaries between the traditional natural fibres of cotton, jute, linen and wool and stimulated new ways of processing and using them. Britain's clothing firms, which are very labour-intensive and number about 7,000 mainly small ones, employ 4 per cent of Britain's manufacturing workforce. Imports of clothing were worth nearly double exports in 1982. Imports of footwear were about four times greater than exports and worth nearly two-thirds of the total output of domestic footwear producers. About half the shoes sold in Britain are made in Britain. Approximately a third of Britain's leather goods (apart from footwear) were exported but imports were worth over 50 per cent more than exports, and they were also worth nearly 50 per cent of Britain's total sales in 1982.

Some points about other manufacturing sectors are worthy of note. Furniture making tends to be restricted to small or medium-sized firms. Exports of furniture have risen in the last decade but so also have imports. The larger owners of paper and board mills have considerable interests abroad, notably in North America, Europe and other Commonwealth countries as well as Canada. Their use of recycled materials has been increasing. Publishing contains a fair number of large firms but in bookbinding, printing, engraving and other associated trades, and indeed elsewhere in publishing, firms are usually small. British book publishing is a major exporter and several of the major firms have subsidiaries in the USA and elsewhere.

Rubber and plastics firms exported a little more by value than they imported in 1982: both imports and exports were each worth nearly a quarter of total sales. About 45 per cent of rubber sales consisted of tyres and tubes, the remainder including conveyor belts, cables, hoses, rubber gloves, clothing and footwear, latex products, and various other vehicle components or accessories. Some production is from recycled and synthetic as well as natural rubber. Plastics are used in myriad ways in manufacturing and construction; they are likely to be used more and more in engineering products, especially in motor vehicles, where their lightness is an asset. About a third of Britain's toy production, where most firms are small, is exported, especially to France, West Germany and the USA. Much recent expansion has been in electronic games and toys. British sports equipment manufacturers are also generally small, but with apparently quite good reputations for quality elsewhere in the EEC and in the USA.

Apart from the scale and the content of the work of the various construction and manufacturing sectors there is the general framework within which they operate. Since the 1960s much of Britain's industrial growth has been associated with offshore oil and gas and with the development and use of electronics and microelectronics. Current government policies aim to reduce government involvement in industry and to support the growth of small firms, hopefully so as to make the economy more competitive. Similar aims underpin measures to encourage employee share ownership, paralleling a growth of co-operative ownership and of staff and/or management company 'buy-outs'.

About 74 per cent of GDP comes from the private sector and the private sector accounts for 72 per cent of total employment. Private ownership predominates in farming, quarrying, mining except coalmining, manufacturing except that of ships and steel, and construction; and in the distributive, financial and various other service sectors. Communications, energy and transport are largely in public hands. In 1984 the private sector employed about 15 million people in about 1.75 million enterprises, of which about 750,000 were corporate (including co-operative) in form. The public corporations, about 16 in number, employed about 1.3 million. They contributed about 9 per cent of GDP; central and local government contributed the other 17 per cent remaining after that and the 74 per cent contributed by the private sector. Most large private sector firms are public companies, which offer shares to the public. Smaller companies are usually private companies, which by law are not allowed to offer shares for sale. The main external sources of finance for company investment are banks, but over half of companies' funds used for investment and related purposes are internally generated. The Stock Exchange provides a great deal of short-term finance. The government, the EEC and specialist financial institutions also help. Recent years have witnessed growth in public and private sector venture capital schemes which offer seedcorn to small and innovative firms, although many commentators have argued that Britain is still a very discouraging place for entrepreneurial types, especially inventors, to work in.

Britons are undoubtedly very aware of the significance of information technology and Britain has been a pioneer in computers and robotics. British skills have generally been greatest in the area of software; they have recently been much less prominent (albeit far from completely absent) in the development of hardware. At the end of 1983 Britain probably remained the world's fifth largest user of robots in spite of a generally poor record in the area and there have been and are several schemes of government support for advanced information technology, fibre optics, optoelectronics, computer-assisted design and manufacturing, and industrial robotics. Biotechnology is another relatively new and very important area, one in which the British record of research and innovation has been excellent. Yet even so Britain has lagged behind Sweden, West Germany and Italy (as well as, of course, Japan and the USA) in the use of robots and Britain's domestic computer industry is currently under threat from abroad, and not for the first time. Although British chemicals manufacturers have a reasonable record of innovation, survival and growth by world standards and a very good record by British ones, there are good reasons for supposing that at least some pioneering efforts in biotechnology will be exploited more successfully elsewhere than in Britain.

The Public Sectors and Public Policies

Government encouragement of manufacturing has taken several other forms apart from the reduction of state ownership and the encouragement of small businesses. Financial and other aids have been given to areas and regions with particular difficulties; there have been several kinds of technical and scientific pump-priming; and selected industries and firms have been given special help. There are several schemes which aim to attract direct investment from abroad; the USA has long invested more in manufacturing in Britain than in any other European country and Japan now invests more in Britain than in all of the rest of the EEC. The taxation system is partly designed to encourage investment.

The main nationalized industries are represented by the British Airports Authority, British Airways, the National Bus Company, British Coal, the electricity generating and supply boards (central and regional), the Post Office, British Rail, British Shipbuilders and British Steel. They employ about 6 per cent of all British employees. British Aerospace, the National Freight Corporation, Britoil, British Telecom, Associated British Ports and parts of British Rail are among the units formerly in the public sector which have been privatized since 1979. Steps have been taken to encourage competition in energy supply, postal services, telecommunications and transport. The top managers of the public corporations are responsible to Ministers although few of them are civil servants. Ministers are supposed only to determine strategy and to refrain from interfering with day-to-day management, but this principle has sometimes been violated. Financial objectives are designed partly to encourage the public corporations to behave as if they were commercial enterprises. Currently they are expected to achieve a pre-tax rate of return on investment of 5 per cent. In general they have increasingly been subjected to financial monitoring and planning. Since 1980 the Monopolies and Mergers Commission has been given powers to investigate their efficiency, and it sometimes uses outside management consultants to that end. Parliamentary committees are also involved in the supervision of public corporations.

For the private as well as for parts of the public sector there are tax and various direct and selective incentives to invest in plant, machinery and buildings. Government research establishments provide various forms of scientific and technical help to industry. The British Technology Group promotes technical change and development by protecting and licensing inventions, funding projects and developments, and by selectively investing in a range of industries. The government also provides technical and managerial advice, credit services and help with training to small manufacturing and service enterprises in rural areas. Northern Ireland, mid-Wales and the Scottish Highlands and Islands all have their own special bodies to encourage development. There are Tourist Boards for and within England, Northern Ireland, Scotland and Wales and for Britain as a whole. The Department of Trade and Industry has a special division which is concerned with policy towards small firms. The Small Firms Service gives advice on over 100 measures which have been introduced since the late 1970s to help small firms.

Government policies to stimulate competition, to control restrictive practices, and to encourage fair trading are the responsibility of the Secretary of State for and the Department of Trade and Industry, and most directly of the Parliamentary Under-Secretary of State for Corporate and Consumer Affairs. The Office of Fair Trading, an autonomous government body, has various duties concerning consumer protection and competition policy. It thus administers Acts of Parliament concerned with Fair Trading (1973, covering monopolies and mergers), Restrictive Trade Practices (1976, restrictive trading agreements), Resale Prices (1976, minimum resale price maintenance), and Competition (1980, other anti-competitive practices). The Monopolies and Mergers Commission deals with monopolies and mergers referred to it by the Secretary of State for Industry and the Director General of Fair Trading. The official definition of a monopoly is quite a broad one which covers both private and public sectors, and national and regional markets. A proposed merger is regarded as meriting public concern if the total value of gross assets to be taken over exceeds £15 million or if a monopolistic situation would be created or increased.

The Office of Fair Trading may look into most kinds of business practice which interfere with competition in the acquisition, production, and selling of goods and services. The public corporations are subject to similar kinds of scrutiny. All restrictive trading agreements, as when two or more firms feel that it is in their interest to limit their freedom to fix prices and conditions of sale, have to be registered with the Office of Fair Trading which refers the agreements to the Restrictive Practices Court: the general effect of this is to curtail and limit the effects of such practices.

The Office of Fair Trading also has powers to protect the interests of consumers, not only indirectly by the above means, but on behalf of consumers as such. It is concerned with consumer affairs in general, with particular trading practices, with consumer credit and hire business, with estate agency work, pricing information and the purity of foods, and with the description and performance of goods and services in general. Citizens Advice Bureaux and Consumer Advice Centres are among the independent agencies concerned with consumer protection. There is also the independent and non-statutory but government-financed National Consumer Council (with associated councils in Scotland, Ulster and Wales), which represents consumers' views to decision-makers in government and industry. The public corporations in energy, rail transport and communications each have their own consumer councils. Several trade associations in commerce and industry have codes of practice. The largest of several private organizations which seek to further consumers' interests is the Consumers Association, with over 600,000 subscribers. It produces numerous very useful publications including *Which?* and its specialist offshoots.

The EEC's consumer programme has acted in several ways as a stimulus to consumer protection in Britain, and the views of Britain's consumer organizations are represented by the Consumers in the European Community Group (UK). Competition rules set out in the Treaty of Rome also form an important part of the backcloth to British competition policy and practice. EEC membership also helps Britain to create jobs and develop the infrastructure and the environment, especially in Northern Ireland and the assisted areas. The European Investment Bank to which all EEC member states contribute, has loaned over £2,500 million to Britain since 1973, for manufacturing projects in the assisted areas and for public works such as the Sullom Voe oil terminal in the Shetlands. The European Coal and Steel Community has given support to redundant British coal and steel workers and to projects designed to create jobs for them. EEC regional policy has two main aims, to reduce regional imbalances and to prevent new ones arising because of EEC policies.

Engineering Employment and Engineering Employers

The national organization of engineering is not only a matter of economic sectors and of relationships with governments. There are also the individual organizational structures within and through which engineers work (see Chapter 11). In Britain these organizational milieux appear to take two major forms, representing two typical and partly contrasting approaches to recruitment to and the structuring of tasks. Across-sector variations in positions of engineers have been important features of this partial contrast.

Engineers' roles in the public sector, often more than elsewhere, have sometimes been limited by a tendency to define engineers as specialists and to give top jobs to arts and natural and social science graduates and accountants. The Civil Service may have been the worst culprit in this respect; in local government at least some top jobs have traditionally been filled by engineers, and in some of the nationalized industries and public corporations such as coalmining and electricity supply and in the 'technological' parts of higher and further education engineers have been very influential indeed.

The 1971 Census showed that over 11 per cent of graduate and nearly 8 per cent of HND/HNC holders were in public administration and defence. The 1983 Engineering Council data showed that 5.0 per cent of Chartered Engineers worked for central government, 10.2 per cent for local authorities, 2.5 per cent for regional authorities, 14.2 per cent for nationalized industries and public corporations, 2.4 per cent for the armed forces and 3.6 per cent for the universities. This and previous CEI data suggest that about 40 per cent of Britain's Chartered Engineers work in the public sector.

In the higher reaches of the civil service, lack of technical knowledge is traditional, and the relevant attitudes have often influenced other large organizations especially in the private sector. However the sectors and units in which engineers' prospects have been much better represent another tradition. Since the nineteenth century there have been two main ways of organizing management in Britain. One pattern, typical of the civil service and many larger companies, is to man top jobs with graduates in traditional academic subjects, because of their lack of vested interest in technical detail, and for their potential for seeing the wood for the trees when important decisions are to be made. They were expected to be able to work in most employment sectors, supported by technical and/or professional specialists such as engineers, accountants and surveyors. This approach was also used in the Colonial Service before the Empire was given away.

The second approach, the opposite of the first, has been more provincial in tone, and its raw material usually came from lower down the social scale, from grammar schools and 'redbrick' universities like those of Reading, Newcastle and Leicester, and from professional offices. In this case technical and/or professional specialists did dominate top managerial positions. In local government, and below the unpaid politicians on councils and their committees who made policies, these were the 'professional empires' of borough engineers, architects, treasurers, surveyors and so on. In construction top jobs are shared amongst civil engineers, surveyors, architects and others, depending largely on the type of work being done and the kind of firm or public sector department.

Most British employing units lie somewhere between the two extremes just described. Since the local government reorganization in England and Wales of 1974 and the development of larger authorities, layers of 'corporate management' have been superimposed over the professional empires, attempting to develop accountants and personnel and other specialists into general or multifunctional managers. In some of the technically and socially less traditional public corporations and nationalized and other industries like aerospace and chemicals, natural science graduates (working by definition as engineers) have become prominent as top job holders.

Since the 1950s those appointed to top posts in industry apart from arts graduates have included accountants, natural science and economics graduates, marketing specialists, and holders of masters' degrees in management subjects. Even in manufacturing the engineer has been relatively scarce at board level as Monck showed three decades ago (Monck, 1954). Sectors in which engineers have been strongly represented at most levels include aerospace, motor vehicles, machine tools, coalmining and electricity supply. The reasons have been quite varied: obviously high-quality technical expertise is central in these sectors; in some of them very strong local engineering traditions have prevailed, and in others, especially the last two, engineers have in effect been long and untypically well represented by professional bodies such as the Institution of Mining Engineers and managerial trade unions such as the British Association of Colliery Managers and the Electrical Power Engineers' Association.

A few other points should be made in this section. First, British private firms, public corporations and government departments tend to be larger on average than is the case in comparable industrial countries. This seems to complicate and exacerbate managerial industrial relations problems (Sorge and Warner, 1980; Prais, 1981a; Child, 1984).

Second, it is worth emphasizing how few engineers in total work in research and development, whereas some journalistic discussions about a meta-world of 'science and technology' often imply that it is a, perhaps the, major category of work for engineers. CEI, Engineering Council and other surveys suggest that it employs only about 10 per cent of Chartered Engineers, and that most of these are in development rather than research. In 1983 a further 2 per cent worked in software development, and nearly 11 per cent in design, and some other Engineering Council categories would include some technical development work (Engineering Council, 1983). The overall research proportion is thus low and the use of 'R and D' as a category seems to be rather status-concerned. Only 3.4 per cent of the chartered engineers in the 1983 Engineering Council survey reported that they worked in production (which seemed however to have been rather narrowly defined). But virtually all survey data give the lie to the popular idea that most engineers work either in production or R and D; their jobs are far more diverse.

Finally we should briefly mention the organizations for engineering employers. Not as well publicized as trade unions, they are nevertheless equally important in industrial relations and they were originally very much in the forefront of the development of collective bargaining. Sometimes – a little misleadingly – called the 'bosses' unions, they negotiate with trade unions in their industries, give advice and services on such matters as pay, work study and employment law to their members, and represent them to government and other public bodies. They are also trade associations insofar as they are often concerned with commercial matters as well as employment ones. A typical representative of an employer on the local, regional or national committee of his employers' association is a senior personnel specialist. Many (especially) private sector employers do not belong to employers' associations: they tend to be either small employers, or ones like Mars and Kodak who do not recognize trade unions, or large ones like ICI, Ford and Chrysler who have strong centralized personnel departments and who cannot be bothered to belong.

The Confederation of British Industries (CBI) is the central umbrella organization for British employers and their associations, the employers' equivalent of the Trades Union Congress (TUC). It was a product of a merger between three separate organizations in 1965, and of a then-growing trend towards bigness in the organization of industrial relations. The CBI has over 4,500 individual member companies, and over 200 employers' and trade organizations representing about 300,000 companies. About 12 million people work for CBI affiliates which, in line with wider trends, are less likely to be in manufacturing than was the case twenty years ago. Altogether there were about 300 employers' associations in Britain in the early 1980s. In the late 1960s there were well over four times as many but many of the smaller local ones have since been wound up or absorbed into regional and national parent bodies.

The largest and most influential body organizing employers in engineering is the Engineering Employers Federation (EEF). Formally established in 1896, it took its present form in the first twenty years of this century. EEF firms and public sector employers tend to be in electrical and mechanical engineering, shipbuilding and aerospace. The EEF has been involved in formal collective bargaining with engineering unions for over 80 years. It is the only employers' association which is not only a trade association. Apart from the EEF there are some sixteen regional or local engineering employers' associations, some very large. In construction there are interesting tensions between associations of local builders and of the larger national contractors, partly because of differences between local and national labour markets and relationships. In recessions, the role of employers' associations in collective bargaining tends to be less important than during better times. Thus when the economy is depressed, individual managements have more power to determine employees' pay and conditions unilaterally and without outside help. The representative and advisory functions of employers' associations are however likely to remain important for the foreseeable future (Sisson, 1983; Clegg, 1979, chapter 3).

Engineers' Experiences of Work: the Data

We have made many references to the experience of engineering in the previous chapters. Here we examine the available data. An interesting study was reported by Faulkner and Wearne (1979). Over 400 practising chartered engineers of all ages were asked about the managerial skills and expertise they needed in their jobs. Just over one in five of the respondents worked in construction; most of the rest worked in manufacturing. Nearly 75 per cent were in charge of two or more people; over 28 per cent agreed that their jobs were predominantly managerial against nearly 22 per cent who reported that they were predominantly technical. Most seemed to have a good deal of say in what work should be done and in how it should be done. Nearly 60 per cent referred to such 'managerial' elements as communications, 'man-management', industrial relations, meeting deadlines, and other administrative, financial and personnel matters as being the most difficult or demanding aspects of their jobs. Only 17 per cent referred to design and other technical matters in the same way.

When respondents were asked to identify their main uses of managerial skills and expertise, the main items mentioned included project and R and D management, 'human relations', health and safety, budget planning and control, personnel selection, industrial relations skills, company law and organization and methods (O and M). Younger respondents seemed to be more concerned with detailed project work; older ones more with man and with some aspects of personnel management. Greater seniority generally meant more concern with corporate, commercial and financial aspects of management. The differences between different types (as well as level) of engineer were often very great: marine engineers were very concerned with health and safety and maintenance, civil and structural engineers with contract management, aeronautical engineers with R and D, and so on. But age and responsibility level were the most powerful predictors, as one might expect, of the managerial emphasis of jobs. Many respondents felt inadequate in terms of 'communications' and 'human relations', in relation to subordinates, superiors and colleagues with other backgrounds. There seem few if any reasons for believing that engineers are any worse in these respects than other specialists such as accountants, or marketing or personnel managers. However, engineers may tend to report any such problems more readily, perhaps as a function of relatively low prospects and status and in some cases their possibly unrealistically poor self-images.

Another recent British study was produced for the Goals of Engineering Education Project for the Department of Education and Science and the Council for National Academic Awards between 1979 and 1983. It attempted 'to evaluate, from an employer base, the products of engineering education' (Beuret and Webb, 1983). The researchers interviewed 250 graduate electrical and electronics and mechanical engineers aged 25 to 35, in 55 public and private organizations selected to represent the national pattern of employing engineers. Another 200 interviews were conducted with (normally more senior) colleagues of the engineers. All of the interviews were concerned with the engineers' backgrounds and attributes and their possession of and need for non-technical abilities.

Both the engineers and their colleagues felt that technical competence was not lacking, but that appreciation of business and especially financial matters were, as were social skills. It was suggested that engineers tended to lack social and intellectual confidence in ways that might damage their career prospects. The engineers felt poorly informed about company policies, only slightly involved in decision-making, and that their companies were poor at understanding technical jargon and at encouraging technical innovation. A self-fulfilling prophecy seemed to operate in many units: engineers were essentially defined as narrow specialists by themselves as well as by their employers. Lack of confidence was associated, in part, with engineering degrees which had a bias towards theory and against practical engineering and its commercial and human realities. Company board members were criticized for being reluctant to consult and promote engineers, a failing attributed to their tendency to come from financial and marketing backgrounds (although non-engineering middle managers in an employing unit could presumably make the same claim). The non-engineers were apparently more likely than the engineers to view their organizations as having a positive attitude to technical development, interdepartmental relationships, innovation and willingness to reward performance with pay and promotion.

A rather low total of 40 per cent of the engineers believed that engineering (18 per cent), production (11 per cent) or R and D (11 per cent) was 'the best career route' in their organization. The engineers were not markedly unhappy with their pay (nor were they exactly happy). A clear majority indicated that they would encourage their children to become engineers; an even larger one asserted that they were proud to call themselves engineers. However 15 per cent thought sales and 29 per cent thought finance were better career routes than engineering in their organizations, the total being greater than the 40 per cent figure noted at the start of this paragraph, and while only 18 per cent clearly had doubts about being proud to call themselves engineers, 33 per cent reported that they would either 'somewhat' or 'strongly' discourage their children from following in their footsteps.

The Beuret and Webb study suggests that younger engineers lack confidence in their own non-technical abilities, and in their employers as users of their skills. Many felt ignorant of their employers' aims and policies and critical, along with their colleagues, of their employers' capacities for innovation. They often regarded their education as having been 'both technically narrow and narrowly technical'. Even the engineering content of courses was often lacking in details which subsequently turned out to be very important in industry. However it is not clear how the details referred to were specifically engineering-related because the neglect was not just restricted to commercial, financial and human topics. From the senior non-engineers, the key criticisms of engineers concerned their 'capacity for technical risk-taking, ability to see engineering in a broader business context, and (understanding of) costing and business finance, and the abilities to explain engineering ideas clearly to non-engineers'. As students, engineers had often neglected courses which were not 'real engineering'; once in industry, they were sometimes 'badly shaken' in realizing how useful they could have been. To sum up, the basic issues were ones of low status, difficulties in recruiting the most able, an education which was relatively narrow, cheap and brief, relatively poor pay and promotion prospects, and a near-total 'absence of engineers from top decision-making levels of the Civil Service, the City, Parliament and the Armed Forces'.

For its first survey of professional engineers the Engineering Council (1983) asked over 22,000 chartered engineers to indicate their degrees of satisfaction with five features of their careers. The percentages reporting themselves as broadly satisfied and broadly dissatisfied respectively with the five features were as follows: initial training (57% satisfied, 14% dissatisfied), training in new techniques (30%, 32%), technical opportunities (46%, 16%), salary and conditions (49%, 20%), and career development (41%, 28%). In each case the other respondents placed themselves in the neutral 'neither satisfied nor dissatisfied' camp. These results suggest that engineers are more concerned about current or recent issues (training in new techniques, career development) than with their initial training – which it is in their interest to praise – or current technical opportunities, which are, perhaps, an important source of satisfaction. In our experience, practising engineers are strongly aware of the fading relevance to tasks of their initial qualifications. They are probably correct to feel this way, as the views of Pym (1969) suggest below, although in doing so they often seem to confuse status-related issues, the difficult but real problems of how to use theory, and their own varying desires for promotion and technical interest and challenge.

There are several earlier studies of experiences in engineering which complement but also in some respects diverge from the more recent findings. An important study of mechanical engineers conducted in the early 1960s did show British engineers to be unhappy with engineering's status in society and – especially in production – within

firms (Gerstl and Hutton, 1966). Dissatisfaction was focused on lack
of power, responsibility and promotion prospects, and on senior
managements' lack of technical understanding (see also Sofer, 1970,
on promotion prospects). Stanic and Pym (1968) also studied
mechanical engineers and 'bureaucratic' stifling of engineering and
engineers' potential. Pym (1969) went on to argue that employers
simply did not know how to reward technical ability, and that
(perhaps for want of anything better) they over-valued educational
qualifications when deciding about pay and promotion. Bamber and
Glover (1975) produced broadly similar findings for technical
specialists working for the British Steel corporation: in a traditionally
engineer-dominated industry, nationalization (in 1967) had led to
strong and often justified fear of 'head-office' planners with backgrounds in finance, marketing and personnel.

Studies conducted in the USA produced comparable evidence.
Ritti (1971) wrote of 'underutilization of skills and lack of organisational influence'. American engineers generally reported lower levels
of job satisfaction than their levels of education and complex jobs
might have been expected to indicate. This was particularly true of
development engineers compared with those in production, sales and
elsewhere. The former often felt that they were being used as glorified
clerks or technicians. Their departments tend to be isolated, tasks
were often highly programmed and job holders were rarely involved
in business decisions. There is almost a vicious circle here in that once
the group comes to be used as glorified clerks the label sticks and
eventually the type of work attracts people who want to be glorified
clerks. Perrucci and Gerstl (1969) and Badawy (1978) argued that
many American corporations were confused about how to employ
graduate engineers. They often failed to appreciate their needs for
responsibility and achievement, and there were many weaknesses in
the provision of technical and managerial training.

Several writers have suggested that the organization of manufacturing units in the USA is often unhelpful to the engineer (as in
Britain). Banks were more prepared to give Japanese (than American)
firms large long-term loans. American executives were too keen to
make short-term profits, at the expense of investment in new processes and products. Law, finance, consulting and industrial relations
were attracting too much talent, engineering not enough. Economic
and social rewards had become detached from genuinely productive
work. Technical innovation and detail were neglected in favour of
'corporate strategies' which involved the over-use of (short-term)
'profit-centres' and over-centralization of research and development
units. More generally, top-heavy and over-complicated 'diversified-

multinational-conglomerate' modes of management exemplified a complacent preoccupation with financial and organizational matters (Glover, 1985).

On the general orientations to work of British engineers, a number of broad concluding points remain to be made. Lawrence (1980) referred to a straightforward competence and respect for detail on the part of West German engineers; similarly Hofstede (1974) suggested that whereas German managers valued clear objectives, orderliness and having authority over others, British counterparts valued achievement and 'challenge'. Both studies imply that British engineers and managers have a less straightforward approach to their work than do Germans (see also Mant, 1978).

Bamber and Glover (1975) found that the main sources of job satisfaction for engineers in British Steel were, in descending order of importance, interesting and challenging work, the exercise of authority, and relationships with others, and that the main source of dissatisfaction was lack of authority, control and status. The first source, interesting and challenging work, was emphasized most by those working in a narrowly defined technical capacity, where the other two more 'managerial' sources of job satisfaction tended more often to be emphasized by engineer–managers and project leaders. This unsurprising finding was also reflected by several of the other studies discussed in this section, although the studies rarely addressed the issue directly (see Whalley, 1986). Bamber and Glover also found that work was less important to engineers than their families or leisure pursuits (see also Sofer, 1970; Child and Macmillan, 1972). Most evidence on the attitudes of all kinds of technically qualified employee suggests that the 'craft' element in their work is very important, and the 'people' element fairly important as sources of job satisfaction. Compared with other specialists with similar levels of qualification, British engineers tend to report fairly low levels of job satisfaction and to tend to be 'privatized' or home-centred along the lines of the affluent manual workers in Luton who we discussed in Chapter 7. The possible exceptions seem to be some of those in research and development whose work can have a more active and innovatory flavour than those of many other engineers (Whalley, 1986).

Finally, in a discussion of the position of engineers in the social structures of advanced industrial societies (especially Britain) Whalley (1986) has expressed strong doubts about a number of theories which – usually implicitly, without much evidence, and in the settings of very general debates – write scripts concerning the roles of technical experts in management and in social and political life. For example some theorists have predicted conflict between engineers'

supposed concern with technical and scientific rationality and professional autonomy on the one hand and employers' pursuit of profits
and control over employees on the other. So-called technocratic
theorists have argued that engineers would inevitably come to control
management hierarchies because of their necessary and superior
technical competence. Other theorists have argued that, as professionals, albeit not very successful ones, engineers would (somehow)
secure occupation-based privileges at work and in labour markets.
More extreme versions of this last thesis have argued that as engineering became more 'science-based' in 'advanced' sectors, engineers'
qualifications became more theoretical and university-based, making
them more independent of employers and (even) leading to situations
in which engineers hire capital (for example by setting up their own
firms) rather than the other way round, or in which they dominate
managements because of the centrality of their inputs to the continued
competitiveness of firms in situations of very rapid technical and
market change. A very different thesis, that of the 'new working
class', sees engineers increasingly regarding themselves as workers,
'not *despite* their technical skills, but *because* of them', because of the
restrictions imposed on them in profit-oriented and bureaucratic
employment situations.

For Whalley there were many faults with these (and other)
positions, ones partly attributable to theorists failing to look hard
enough for evidence and to a tendency to generalize too much across
societies without considering often very important differences
between them. He characterized British engineers as 'trusted
workers' whom employers allow a fair amount of freedom, status and
other privileges compared with most other employees. This was
because it was not in the interest of employers to treat them as ordinary
workers when they were capable of forming 'part of the pool from
which further high-level managers are to be selected'.

Engineers in Britain were broadly 'part of management', less
obviously and realistically so than French counterparts for example,
but with many interests in common with other 'senior' managerial-
level employers nonetheless. Engineering knowledge was not a threat
to managements; it was not anything like as 'autonomous' and
'science-based' as some academic observers assumed. It was essentially technical knowledge, subservient to the employers who paid for
it. It was superior to the knowledge of craftsmen and production
workers partly in the sense that it was sanctioned by employers ' in a
package also containing "trustworthiness" and "responsibility".' In
firms with a very strong need for creative technical ability, engineers
were generally given a lot of freedom, although insulated from com-

mercial management so as to forestall possible conflicts between their technical interests and their employers' commercial ones. In more 'conventional', less innovative firms, engineers' interests tended to be directed more towards managerial involvement. Here, commercial considerations were usually brought into play to show engineers that any constraints on their work or prospects were due to 'objective' market forces, not to 'the whim of management or the search for capital accumulation'.

Conclusions

This chapter is expressly about engineering work; and, while we are the first to accept that this is very varied, we are still conscious that it is necessary to summarize what this typically comprises. All generalization is linked with ideas of what is typical; and ours forms no exception to that rule.

First, we must stress that the typical work of engineers, and the typical work involved in sectors of employment known as 'engineering', is concerned with the making and use of concrete objects, many of them made of metal – machines, some building structures, means of transport – and almost all of them bought by users for their utility. Second, because these objects are typically complex, the maker of them has to be competent in the assembly of various components, themselves often fashioned, manufactured products, which have to fit together suitably: the user–engineer has to be skilful and knowledgeable to be able to understand how this assembly takes place and how it is designed to work. Third, although the engineer should be familiar with the requirements and characteristics of the users of his or her products – such as (say) in the design of machine-tools – the special expertise involved centres on the use and making of 'engines'.

In this conception of engineering work, the practitioner should certainly be familiar with the principal known characteristics of the materials which make up his or her product, as are available from scientific subject areas such as physics, chemistry, metallurgy and so on. But it is apparent that access to such knowledge, and understanding of it, forms only a minor part of overall specialist competence, as is indicated by the use of the English words, artefact, artifice and art in this context.

'Art' is to do with jointing individual items or components together so that the whole assembly operates adequately according to certain design-criteria for product- and artefact-performance; and the various parts of (natural) science have virtually nothing to say on this at all. Furthermore, with the engineer seen thus as an artist of a type, something loosely known as 'engineering science', or perhaps 'technological knowledge', cannot possibly have the determining degree in technical and engineering effort that is sometimes claimed for it. This is because of the inherent uniqueness of technical problems and technical solutions, given the nature of the work undertaken by engineers and their immediate colleagues in fashioning and putting together components of 'engines'.

Engineers are not hired primarily because they have read 'the book', or because they own a copy of 'the book' and can understand what its author(s) say(s); rather they are used, and valued, because of their ability to work out solutions on matters for which 'the book' says little or nothing.

This is the core of the engineering task. It is the meaning which the engineer ascribes to this task that to a large degree determines the nature of the experience of engineering. The questionnaire data reported in the previous section do not perhaps make this point squarely enough in the sense that other variables associated with status, prestige, power, social backgrounds and so on have rarely been introduced to the analysis. The action frame of reference's emphasis on the social construction of reality tends to be overlooked. Nevertheless the basic conclusion which we draw is that although sometimes excluded from or simply discouraged with regard to management, many British engineers console themselves with interesting technical work and their home lives. The relatively marginal status of engineering in British employing units does not usually produce major problems at the level of the individual and individual engineers live relatively comfortably with their position. Interest in the job at hand has collectively helped engineers to tolerate their poor status position and their constrained involvement in management. However at the societal level the marginality of engineering may indeed have profound consequences, not for individual engineers, but rather for society as a whole. This is our focus in Chapter 11. In our next chapter we consider the collective responses which engineers have made to their position.

10 Engineers, Professional Associations and Trade Unions

Introduction

In Britain two types of organization seek to influence opinions and to pursue particular interests – political parties and pressure groups. Political parties, whose aims are to secure political power and make use of this in various ways (in local or national bodies), have existed in something like their present form in the United Kingdom since the second half of the nineteenth century. Pressure groups predate political parties. Their main purpose is not to seek and use political power but to pursue some special interest by persuading those in power, directly or indirectly, to alter policies in some way favourable to their particular causes. There are thousands of pressure groups in Britain representing a vast array of particular views and interests. Some are permanent like the Royal Society for the Prevention of Cruelty to Animals, the National Society for the Prevention of Cruelty to Children, the Campaign for Homosexual Equality, the Campaign for Nuclear Disarmament, and the Society for the Protection of Unborn Children. Other pressure groups exist only temporarily to pursue a specific aim like opposition to the building of a motorway, power station or airport, and once either the aim is achieved or they are unsuccessful they disband. There is also a highly specialized type of pressure group whose activities are permanent and who are distinguished from other types because they represent particular occupational groups. There are two different varieties of this kind of group: professional associations and trade unions (Blondel, 1967; Brown, 1981; Clegg, 1979; Johnson, 1972).

In this chapter we examine the way in which engineers have attempted to both serve the public interest and press their own special interests through professional associations and trade unions. Neither the professional bodies nor the unions have been particularly successful in representing the special needs and interests of engineers in Britain or in giving them a voice in economic bargaining. This chapter considers the reasons for the relative failure of engineers' collective organization in Britain. It is argued that a notable feature of this failure has been engineers' preference for professional associations rather than trade unions, although we do not argue that the latter offer the solution to engineers' problems. The underlying reasons for the preference and why this strategy has ultimately been unsuccessful are also analysed.

Professional Bodies and Trade Unions Representing Engineers

Both the history and contemporary picture of the collective organization of engineers are characterized by fragmentation. The first engineering association appeared in 1771 (the Society of Civil Engineers) but it was an informal, learned society for a few senior engineers rather than a representative or qualifying body, and other organizations soon developed. These in turn gave birth to still more institutions. The Institution of Civil Engineers was formed in 1818, and obtained its Royal Charter in 1828. Until the 1870s it was primarily a study society, concerned with exchanging ideas and advancing knowledge. It developed an interest in education and training when, in the middle of the nineteenth century, facilities for engineering education were established in universities and elsewhere. In 1897 it established its own examinations. The Institution of Mechanical Engineers was founded in 1847 and received its Royal Charter in 1930. The Institution of Electrical Engineers appeared in 1871 gaining its Royal Charter in 1921. The Institution of Production Engineers was founded in 1921 partly as an offshoot of the Mechanicals.

There are now sixteen chartered and at least thirty-seven non-chartered institutions for engineers in the United Kingdom. The chartered institutions were members of the Council of Engineering Institutions (CEI) established in 1965 to co-ordinate activities and to represent the engineering profession as a whole. In response to many of the criticisms made of the representation of engineers and following the recommendations of the Finniston Committee, the Engineering Council was established under its own Royal Charter in 1981. The CEI ceased to exist in 1983. The principal aim of the Council is 'to advance education in, and to promote the science and practice of, engineering for the nation's benefit and to promote industry and commerce in the United Kingdom'. Additionally the Engineering Council attempts to co-ordinate educational and learning activities of engineers by making recommendations to the engineering institutions, the universities and validating bodies about the nature of educational qualifications and procedures for gaining professional status. The Engineering Council also aims to provide a unified voice for engineers to government, industry and the educational institutions.

The non-chartered institutions tend to be incorporated as companies under the Companies Act. The chartered and non-chartered institutions together represent over forty types of practising engineer in Britain. (A full list of associations is given in Table 10.1 opposite.)

The CEI did not and the Engineering Council and the Institutions do not engage in collective wage bargaining. The Institutions aim to be learned bodies concerned with education and qualifications and under the terms of their Royal Charters do not interest themselves directly in salaries and conditions of employment. In 1939 a body known as the Engineers' Guild was established. Its Memorandum of Association stopped it supporting trade union activity. However the Guild did attempt to further the interests of engineers by offering services to individuals and by publicly campaigning on behalf of engineers. The aim of the guild was to represent engineers as a profession in the same way that the Law Society and the British Medical Association represent the legal and medical professions (Prandy, 1965).

There are a number of trade unions representing practising engineers although this is a twentieth-century phenomenon. Unlike the professional institutions, the unions representing engineers *do* engage in direct collective bargaining. Whether engineers join unions (and only a minority do) depends upon employee attitudes, employer attitudes, the size of the firm, the hierarchical and bureaucratic nature of the employing organization, and the presence or absence of unions willing to recruit them (cf. Bain, 1970).

A number of different trade unions represent British engineers. The United Kingdom Association of Professional Engineers (UKAPE) was founded and registered as a trade union in 1969. It was started by members of the Engineers' Guild with which it shared accommodation. Its philosophy was avowedly non-political – by which it meant it shunned involvement with organized labour and other unions in the TUC, supported the Conservative Party, and rejected the idea of militant action as unsuitable for 'professional' employees. It was largely unsuccessful due to its elitism, hostility from other unions, and its lack of strong roots in any industry which meant that employers were unwilling to recognize it (cf. Dickens, 1972, 1975) and it was absorbed into a bigger union in 1980. However a number of other trade unions which do have links with the broader labour movement also represent engineers. Recent figures suggest that some 41 per cent of Chartered Engineers and 54 per cent of Registered Technician Engineers belong to trade unions. Membership concentration is higher in the public than the private sector and young or

Table 10.1   *The Engineering Institutions, Associations and Societies,
1983*

Royal Aeronautical Society
Institute of Energy
Institution of Chemical Engineers
Institution of Civil Engineers
Institution of Electrical Engineers
Institution of Electronic and Radio Engineers
Institution of Gas Engineers
Institute of Marine Engineers
Institution of Mechanical Engineers
Institution of Metallurgists
Institution of Mining Engineers
Institution of Mining and Metallurgy
Institution of Municipal Engineers
Royal Institution of Naval Architects
Institution of Production Engineers
Institution of Structural Engineers

> The chartered institutions, formerly members of the Council of Engineering Institutions

Association of Mining, Electrical and Mechanical
    Engineers
Biological Engineering Society
British Institute of Non-Destructive Testing
Institution of Agricultural Engineers
Institution of Highways and Transportation
Institution of Hospital Engineering
Institution of Nuclear Engineers
Institution of Plant Engineers
Institution of Public Health Engineers
Institution of Public Lighting Engineers
North East Coast Institution of Engineers and
    Shipbuilders
Welding Institute

> Formerly Council of Engineering Institutions Affiliate Members

Association of Water Officers
Bureau of Engineer Surveyors
Chartered Institution of Building Services
Highway and Traffic Technicians Association
Institute of Automotive Engineers Assessors
Institute of Engineers and Technicians
Institute of Measurement and Control
Institute of Metallurgical Technicians
Institute of the Motor Industry
Institute of Plumbing
Institute of Quality Assurance
Institute of Road Transport Engineers
Institute of Sheet Metal Engineering

> Other bodies

continued

Institution of Electrical and Electronics Incorporated
   Engineers
Institution of Engineering Designers
Institution of Engineers and Shipbuilders in
   Scotland
Institution of Mechanical and General Technician
   Engineers
Institution of Railway Signal Engineers
Institution of Technician Engineers in Mechanical
   Engineering
Institution of Works and Highways Technican Engineers
Minerals Engineering Society
Society of Civil Engineering Technicians
Society of Electronic and Radio Technicians
Society of Licensed Aircraft Engineers and
   Technologists
Society of X-Ray Technology

                                    Other bodies

Sources: The 1983 Survey of Professional Engineers: A Survey of Chartered and Technician Engineers, The Engineering Council, 1983, pp. 41–2, and *The Engineers Registration Board*, The Council of Engineering Institutions, 1982, pp. 4–9.

recently qualified engineers tend to be more favourable towards joining unions than older men (see Tables 10.2 and 10.3).

   Data collected by the CEI suggests that membership of trade unions among qualified engineers has been slowly increasing over most of the last 20 or so years, although the evidence for the last few years is obscure and the trend may have been halted or reversed. This may have been because, as the Finniston Committee noted, the professional bodies are themselves now more willing to recognize the rights of qualified engineers to join trade unions than they once were. However, growth in public, compared with private, sector employment of engineers was probably a more basic factor; public sector employers have traditionally been far less antagonistic towards trade unions than employers in private industry (Bain, 1970).

   There is, then, a large number of professional associations and unions which represent engineers in Britain. Yet far from such an apparently extensive coverage producing effective representation and powerful pressure groups to speak on behalf of engineers, the opposite seems to be the case. Since the Second World War the status of engineers in Britain has not only not improved much but has probably declined (cf. Watson, 1976). In the same period engineers have tended to maintain their living standards compared with manual

Table 10.2   *Trade Union Membership of Chartered Engineers*

Name of Trade Union	Trade Union Initials	Trade Union Members %	Chartered Engineers %
National and Local Government Officers' Association	NALGO	20.6	8.4
Institution of Professional Civil Servants	IPCS	11.0	4.4
Engineers' and Managers' Association	EMA	10.0	4.1
Association of Scientific Technical and Managerial Staffs	ASTMS	8.6	3.5
Electrical and Engineering Staff Associations	EESA	8.4	3.4
Amalgamated Union of Engineering Workers – Technical Administrative and Supervisory Section	AUEW– TASS	6.7	2.7
Association of University Teachers	AUT	6.3	2.6
National Association of Teachers in Further and Higher Education	NATFHE	4.6	1.9
Society of Post Office Executives	SPOE	2.9	1.2
Association of Public Service Professional Engineers	APSPE	2.9	1.2
Transport Salaried Staffs' Association	TSSA	2.4	1.0
British Association of Colliery Management	BACM	2.3	0.9
Merchant Navy and Airline Officers' Association	MNAOA	1.4	0.6
Association of Management and Professional Staffs	AMPS	1.3	0.5
Association of Broadcasting and Allied Staffs	ABS	1.0	0.4
Association of Polytechnic Teachers	APT	0.9	0.4
Greater London Council Staff Association	GLCSA	0.4	0.2
Other or not specified		8.3	3.3
Total		100.0	40.7

Source: The 1983 Survey of Professional Engineers: A Survey of Chartered and Technician Engineers, The Engineering Council, 1983, p. 14.

workers but when compared with professionals like doctors, lawyers and solicitors, and with senior management in industry, their living standards have fallen behind (Routh, 1980).

In the remaining sections of the chapter an explanation for this state of affairs is advanced. The explanation falls into three parts: first, a consideration of why the professional type of organization and association has been relatively attractive to many British engineers; second, a consideration of the extent to which the professional engineering organizations correspond to the ideals of a true profession; and third, a review of why engineers have been lukewarm towards trade unions.

Table 10.3   *Trade Union Membership: Technician Engineers*

Name of Trade Union	Trade Union Initials	Trade Union Members %	Technician Engineers %
National and Local Government Officers' Association	NALGO	22.7	12.3
Association of Scientific Technical and Managerial Staffs	ASTMS	14.5	7.9
Amalgamated Union of Engineering Workers – Technical Administrative and Supervisory Section	AUEW– TASS	10.2	5.5
Institution of Professional Civil Servants	IPCS	10.1	5.5
Engineers' and Managers' Association	EMA	9.6	5.2
National Association of Teachers in Further and Higher Education	NATFHE	9.2	5.0
Society of Post Office Executives	SPOE	2.0	1.1
Association of Broadcasting and Allied Staffs	ABS	1.0	0.5
Transport Salaried Staffs' Association	TSSA	1.4	0.7
Merchant Navy and Airline Officers' Association	MNAOA	0.8	0.4
Other or not specified		18.5	10.1
Total		100.0	54.2

Source: The 1983 Survey of Professional Engineers: A Survey of Chartered and Technician Engineers, The Engineering Council, 1983, p. 14.

The Attraction of the Professional Ideal

The reason for the attraction of professional organizations for many (although not all) engineers lies in the origins of the professional ideal itself. Two medieval institutions were the antecedents of the professions in Britain – the guilds and the Church. The English guilds can be distinguished into two – the guild merchants and the craft guilds. The guild merchants were groups of merchants who protected their own economic interests in particular towns. They inspected markets, judged the quality of merchandise and laid down rules of business and manners. In other words they were self-regulating bodies. The craft guilds represented skilled tradesmen and regulated working standards, labour prices and the apprenticeship system. In some crafts a seven-year apprenticeship was thought essential for the skills or 'mystery' of a craft to be acquired. Sometimes the regulations of both types of guild were enshrined in statutes or byelaws. The role of the state, or rather its handing the role to others, is important here. The state allowed the guilds to undertake certain regulatory functions in lieu of state activity. The other main influence on the professional ideal was the Church with its emphasis on altruism, its role as mediator between God and man, and its virtual monopoly over education, literacy and numeracy. The medieval Church encouraged, not to say demanded, the faith and obedience of its members and a belief among the congregation that the Church knew best. The faithful were promised salvation so long as they followed the regime laid down by the Church. In medieval Christendom the Church offered a hope of salvation not obtainable by secular means. The most important feature in the context of our argument which these three groups shared was their ability to dominate the relationship between themselves and those people for whom they provided services, secular or sacred.

During the eighteenth and nineteenth centuries the elements of restriction of practice through self-regulation and protection of self-interest (the merchant guilds), restriction of entry and control of training (the craft guilds), and control of specialized knowledge, altruism and mediation between the sacred and the profane, along with the ability to dominate client groups, fused together in the formation of the medical profession in Europe. Subsequently the type of organization (the profession) developed by doctors to maintain these ideals became a model of collective organization which has been widely imitated and copied.

The medical profession and with it modern medicine and medical practice was a hybrid of three distinct occupational groups who practised in the early eighteenth century. There were first the physicians (doctors) who learned their medicine from textbooks, usually with no practical or clinical experience. By modern medical standards the textbook knowledge was nonsensical. However the physicians were prevented from doing very much harm because the patients dominated the relationship between practitioner and patient. The patients tended to be the rich or well-to-do and were thus able to resist medical diagnosis and treatment if they felt that it would be harmful. A physician, like a clergyman, was valued for his wit, charm and bedside manner as well as his ability to mix socially with patients. In practice medical expertise was not the main criterion by which a doctor was judged. Of course even today bedside manner and a pleasant personality are liked by patients. The second (and then the definitely inferior) group were the surgeons or barber–surgeons. To this day their old lower status is perpetuated in that surgeons are called

Mister while a physician is Doctor regardless of qualification or experience. The surgeons, who had their own guild, were practitioners of applied anatomy learned on the job. The Royal Navy in particular was a major source of employment because sailors wounded either in battle or everyday seafaring required immediate life-saving amputations or other procedures – and not witty bedside chat. The surgeon was therefore judged on the efficacy of his physical interventions. The third group were the apothecaries or druggists who sold herbal and drug cures over the counter in their pharmacies and shops. These were tradesmen and shopkeepers and considered the lowest status group among those offering medical services.

The partial unification of the three groups into a single professional grouping followed in the wake of growing systematic research and experimentation on patients which originally occurred in France following the revolution there and the subsequent nationalization of the Paris hospitals. In post-revolutionary France research in anatomy, physiology and pathology flourished and the results were transmitted in the newly founded medical journals of the period. In Britain these developments were followed with interest and a unified medical profession developed to practise medicine. The new medical profession consisting of physicians and surgeons established a state-regulated monopoly of medical practice. Their practice enshrines restriction of entry, self-regulation, control over practice and an increasingly specialized knowledge.

The notion of 'knowing best' and the encouraging of faith in the powers of the medical profession were directly borrowed from the Church. The sick were encouraged to put their faith in the doctors who 'knew best', and in return would receive a cure, or relief of pain and suffering. It was not the patient's place to question the nature of the therapy offered. A good patient followed doctor's orders as a matter of faith, a belief that the doctor knew what he was doing.

In return for allowing the monopoly (which did meet some resistance) the state got a gradually more expert medical service – although the improvements in health (measured in terms of a falling death rate) which occurred in the late nineteenth century owed more to the engineers who built sewers and provided clean water and so prevented much disease in the first place than to the growth in specialized medical knowledge or the monopoly of medical practice (Holloway, 1964; Waddington, 1973).

The high status enjoyed by the medical profession is a result of three factors. The profession took on a highly respectable aspect – a legacy from the old physicians and their aristocratic patients. The medical profession had access to mysterious special knowledge which was not available except to those admitted to medical training. Finally the profession had a mystical quality derived from the healer–sufferer relationship. In almost all human societies, healers (be they witch-doctors, shaman or medicine men) hold a special or venerated position in society. The reason for this is not difficult to discern – the sufferer, who is ill, in pain or dying, requires consoling both spiritually and temporally. Persons providing such consolation hold much power. In this sense the mystical side of medical practice was derived from the priestly activities of the Church and its confessional and communion rituals as well as its role in caring for the sick in monastic settings in the early medieval period.

The venerated position and the scholarly and respectable ambience of the medical profession has attracted many imitators. In particular the expanding middle classes of eighteenth- and especially nineteenth-century Britain provided a fertile breeding ground for aspiring professions of every description (Reader, 1966). The conditions which caused the growth of the middle classes included first the increasing complexity of agriculture, trade and commerce particularly as statutory intervention multiplied (e.g. the Enclosure, Companies and Factories Acts) which demanded financial and legal experts like surveyors, lawyers and accountants. Second there was the development of new industries (e.g. chemicals and foodstuffs) which required technical and scientific knowledge and expertise. And third there came the growth of state intervention in a vast number of activities including health and welfare, education and the nationalized industries, as well as in the expanding local and national government. Occupational groups in these new or expanding fields like accountants, nurses, scientific civil servants, hospital administrators, treasurers in local government, teachers and engineers, and a multitude of others in the private and public sectors have all attempted to adopt professional modes of practice and organization.

Is Engineering a True Profession?

Of the many groups aspiring to professional status and prestige some have been more successful than others in achieving it. The degree to which groups may be considered to have been successful may be determined in part by considering what the defining criteria of a profession are. Unfortunately this is no easy task since commentators have failed to agree on a single definition of professions. One group of writers has argued that the most important characteristics on which to base the definition are: specialist knowledge, self-regulation, and the sense of honour and prestige conferred by membership. Other writers have also included altruistic service, the possession of a code of ethics, high financial rewards, guarantees of competence for the client (given by the possession of degrees, diplomas and certificates) and the existence of a professional body. On a more sceptical note other writers have argued that a profession can be distinguished from other occupational groups by its monopolistic restrictive practices, the artificially high prices for services provided, the absence of effective public scrutiny of professional behaviour and the ability to bamboozle the public.

In a rather different vein it has been suggested that the key to understanding the true nature of a profession lies in the form of collective organization, arising out of the context of the relationship between the professional and the person to whom the professional's service is given. It is argued that the fact which distinguishes the relationship between a professional and his or her client from other kinds of relationship, financial or service, is that the client is in a very weak position to judge the nature of the service provided, at least in the short run. When a doctor issues instructions concerning a drug regime or recommends surgery, for example, the patient does not possess the technical expertise to judge the accuracy of the doctor's prescription. In the long run commonsense criteria may be used by the patient to evaluate the treatment – the patient will know whether or not he or she feels any better. In the short run however the patient has to take the advice or the medicine on trust. It is argued that in order to protect the client in this or other kinds of similar relationship various kinds of control over professional practice have been instituted. (It should be noted that in using this approach some occupations not traditionally thought of as professions get included in the definition.) First there is colleague control, whereby the occupation is self-controlled with no or minimal outside interference. Medicine and law are the best examples of this type of organization. Second there is patronage, whereby the client directly employs the professional either sporadically or permanently. Accounting in various forms offers a good example of this. Third, there is mediative or third-party control, where some outside agency (usually the state) controls the group's activity. Teaching is a good example of this (Johnson, 1972).

Some occupational groups have been particularly successful in organizing themselves around all or some of the principles outlined here. In Britain chartered accountancy stands out as a good example of how the lowly job of book-keeping has become elevated to 'a management decision-making specialism' complete with all the professional trappings. On the other hand the attempts by nurses to claim to be a profession have been far less successful.

Whether or not a group is successful in its claims to be a profession in the British and North American context stems from its ability to dominate the client group. For the medical profession, domination over the client is a relatively easy matter since either the patient is not in a position to resist or because for many (perhaps the majority of) patients the doctors' remedies plainly work or at least relieve pain and suffering. For accountants and lawyers the ability to dominate the client is assured because their skills lie in disentangling a complex web of laws and rules. The complexity puts the client at a distinct disadvantage and very often the only guarantee the client has that the service he receives is up to scratch is the fact that the accountant or lawyer is a member of one of the relevant professional bodies.

Using the various criteria here described it is possible to consider the position of engineering as a profession. Engineering presents a mixed picture. Engineering has a number of specialist knowledge bases. But more importantly engineers also possess the skills of making and doing which are related to the knowledge base. While they, like any other profession, cannot stop people examining their knowledge base (it is available in engineering textbooks) the skill of putting it to work is clearly not available to the public at large. In the long run their activities are subject to commonsense evaluation just like those of the doctor – 'does something work?' is the same as the idea 'do I feel better?' after following my doctor's advice. As with the doctor it is the short-run judgements which are difficult for the client to evaluate.

In terms of self-regulation through a professional body it is difficult to see how the mass of sometimes competing engineering institutions qualifies, with their differing standards and membership requirements not to mention their lack of coverage of all engineers. Even more important, whereas in medicine clinical decisions and clinical practice are controlled almost exclusively by the medical profession – the profession is self-regulated – in engineering it is non-engineers (employers, contractors, architects, accountants, marketing directors) who at least partly define the problems and to some extent the solutions. However if one thinks of the various third parties which mediate between engineer and client like the British Standards Authority, the Health and Safety Executive, and of numerous other state controls, then a good case can be made for this particular version of the professional model applying to engineering.

Honour and prestige are traits frequently associated with professions. All the evidence suggests that in Britain most engineers enjoy neither great honour nor great prestige. Whatever high status engineers enjoyed in Britain in the nineteenth century has to a large extent disappeared (Watson, 1976). The codes of ethics and doctrines of altruistic service which have been said to characterize professions are not obvious in action as far as most engineering is concerned. To many, the engineer is merely a skilled person who sells his abilities to the highest bidder in the pursuit of self-interest. The skills are used in the building or making of things – some of which may well be in the public interest, others which may not be.

Many but not all engineers are in a situation of patronage in so far as they are controlled by a paymaster. Unlike accountants in private practice, however, whose relationship with clients already is one of patronage, almost all employed engineers' skills are not statutorily defined and therefore engineers cannot enjoy the autonomy which accrues to such accountants. Thus in the final analysis engineers are rarely in a position to dominate clients. They are mainly employees who sell their skills. In such circumstances to refer to engineering as a profession befogs the issue. For many engineers probably the best description of them is that they are highly qualified and skilled salaried employees. Indeed this seems to accord with the engineers' own self-images and behaviour in so far as only about half of those eligible to join the relevant professional engineering bodies in Britain do so. Moreover at least some of those who do belong have lukewarm attitudes towards them, seeing them as sources of slightly enhanced qualifications and as ways of keeping in touch with wider developments which are on the whole more interesting than useful (Bamber and Glover, 1975).

The whole discussion of professions has to some extent been a very culture-specific one. We have taken the case, probably the very special case, of medicine and measured other occupations against it in terms of various criteria which have been called professional. The point, however, is that the most interesting feature of this whole discussion is that the engineering institutions take the professional ideal so seriously. The reason that they do is rooted in the fact that the professional ideal has a very powerful status value in British and other major English-speaking societies. British engineering institutions are doing nothing more than showing their characteristic Britishness. The professional ideal is far less strong in the non-Anglo-Saxon world where different specialist occupations have been developed in response to technical change under state control and where self-governing, self-qualifying occupations tend to be conspicuous by their absence.

Thus the lack of attractiveness of the professional ideal in conti-

nental Europe is also historical in that the potential development of
senior occupations into professions was effectively stifled mainly in
the nineteenth century. In Britain, in contrast, in the atmosphere of
laissez-faire, professions flourished and proliferated with state
approval (see Buchanan, 1985, on engineering).

Engineers and Trade Unions

The question which now arises is why have engineers been equally
lukewarm to the trade union movement, given that they are in an open
labour market selling their skills and given that collective organization through trade unions has been traditionally the way in which
skilled people have attempted to manage the selling of their skills in
that situation? The answer to the question lies in two partly distinct
aspects of British trade-union history – first its origins among manual
workers and second its relatively late extension to non-manual
groups.

As with the professions there has been some argument in the
literature about how to define a trade union. Legal definitions
enshrined in statutes are generally agreed by social scientists to be of
use only in a narrow legal sense, mainly because associations of
employers have been included in legal definitions. Of course
employers' associations are pressure groups but not trade unions in
both the commonsense and technical sociological senses of that term.
For the purposes of this chapter a trade union is a protective
association of a collectivity of employees whose strength is greater
than that of its individual members. The primary function of a trade
union is the protection of the living standards of its members, mainly
as measured by income or job protection. Trade unions also pursue
other functions with varying degrees of enthusiasm, including
involvement in party politics, providing friendly society benefits for
members such as health, sickness and death benefits, and providing
educational services. Trade unions also act as pressure groups above
and beyond their purely occupational interests and may support or
attack a whole range of positions or ideas which strictly speaking have
nothing to do with employment matters (Clegg, 1979).

There are numerous arguments about the origins of trade
unionism. It has been suggested that trade unions are the modern
descendants of the medieval craft guilds. This view has been hotly
disputed by those writers who have argued that when trade unions (in
Britain at least) emerged they were quite distinct from any earlier type
of organization. The merits of either argument are a matter of

historical debate. What is clear is that the particular form of organization here defined as a trade union emerges in Britain at the same time as society changed from an agrarian to an industrial one. In fact trade unions have grown up in all Western industrial societies when and where the size of employing unit has increased, where working people have either sensed their own exploitation or had others articulate it for them and attempted to defend their own particular interests. On the whole, the more economically developed the capitalist industrial society the greater the freedom which unions have enjoyed and the greater the extent to which society has been able to accommodate conflict between labour and capital (Kerr *et al.*, 1973; Dahrendorf, 1959). Trade unions were both a consequence of and a response to industrialism.

In Britain the conventional wisdom about the origins and history of trade unions stems from a study by two socialist intellectuals Sydney and Beatrice Webb (1920). They argued that the development of trade unionism can be divided into a number of distinct historical stages. First in the years prior to 1829, which the Webbs referred to as the struggle for survival, nascent craftsmen's groups and other embryonic trade unions fought to maintain their existence in the face of hostility from employers, the judiciary and parliament. They argued that this period came to an end with the repeal of the Combination Acts which had made it illegal for groups to have any kind of collective organization. The next stage (1829–42) the Webbs called the revolutionary period. This was when working people began to engage in mass action and when one large and potentially very powerful union – the Grand National Consolidated Trade Union (GNCTU) – was formed.

There then followed the period of 'model unionism' (1842–75). This was when small craft-based unions developed which had exclusive terms of membership (the possession of a particular skill) and high rates of subscription. These associations were characterized by a pragmatic and piecemeal approach to their business with a strong emphasis on friendly society benefits. The model unions stressed the virtues of hard work, sobriety and self-help and were thus typical of much upper- and middle-class Victorian thinking. The Webbs have emphasized that the model of organization established by these unions was very important in determining the future development of trade unionism in Britain – conservative in outlook and defensive in orientation. These unions defended their members' living standards not only from wage-cutting employers but also from unskilled and semi-skilled workers who might attempt to undercut them in the labour market. The period of model unionism drew to a close with the passing of legislation in 1871 and 1875 which effectively guaranteed the legality of all trade unions.

The next period identified by the Webbs is the period of 'new unionism'. This era (1875–90) saw the growth of mass unions which had larger and less skilled memberships than the model unions but which by virtue of their size were able to wield considerable power. Two important strikes are conventionally identified as of major importance in the development of new unionism – the Bryant and May matchgirls' strike of 1888 and the 'dockers' tanner' strike of the following year. In both cases hitherto unorganized groups of unskilled workers struck very successfully against powerful employers. It is usually argued that general unionism in Britain today may be traced to the new unionism of the 1880s.

The final period identified by the Webbs (1890–1920) was characterized by steady growth in union membership, but more importantly by the development of the party political wing of the union movement. Political representation for trade unionists became a necessity following a number of judicial rulings which stripped trade unions of the power and legality they had gained in the 1870s. In one such judgement, called the Taff Vale decision, it was ruled that a trade union was liable for damages to an employer as a consequence of industrial action. Out of the campaign to defend trade unions the Labour Party eventually emerged whose aim was to represent the interests of the trade unions in parliament. From that time the trade unions have been the principal funders of the Labour Party.

Since 1920 there have been further developments in the structure of trade union representation in Great Britain. There has been a general but fluctuating increase in the numbers of union members. At the same time unions have been subject to various legal controls. More important, the amalgamations and changes in the structure of employment have to a large degree affected the shape of the trade union movement. The present structure of British unionism defies easy classification with some unions based on occupations, others on economic sectors or parts of them, some open to numerous occupations in a variety of industries and so on (Jackson, 1977 and 1982).

British unionism acquired its traditional character – manual and working-class, socialist in principle but conservative in orientation and action – in its formative period. This gives us a clue as to why the early manual unions offered nothing to the most highly qualified professional engineers even though many of these engineers often worked in close proximity to manual employees and craftsmen engineers were in the forefront of model unionism. The early unions were not interested on the whole in the professional middle classes and the feelings were undoubtedly reciprocated. However this does not explain why the unions which developed to represent non-manual employees should also have proved so unattractive to engineers, except in the present century and in parts of the public sector.

The numbers of persons engaged in non-manual work has increased both absolutely and relatively since the middle of the nineteenth century. Since then an ever-smaller proportion of the working population has been engaged in the extraction of basic food-stuffs and raw materials from the natural environment and a greater proportion have been employed in manufacturing industry and in the commercial, business, administrative and service sectors. All of these sectors, apart perhaps from agriculture, but certainly including the extractive and manufacturing sectors, employ large numbers of non-manual people (Bain, 1970; Lockwood, 1958).

While non-manual occupations have expanded and the potential number of non-manual union members has also increased, actual union membership among these workers has not advanced at the same rate as the potential. Many commentators have noted the fact that many white-collar workers, including engineers, appear to be resistant to trade unionism. It is worth remembering, however, that even though the rates of unionism among non-manual workers are generally low, some non-manual workers such as Post Office workers, teachers and civil servants, have had a long history of trade union organization and high levels of union membership. As we have noted non-manual workers are generally much more highly unionized in the public than private sectors, and among them men are more highly unionized than women. Also white-collar unions have been less willing than their manual counterparts to go on strike (although in the 1970s and 1980s several groups of white-collar workers, notably civil servants and teachers, did engage in prolonged strike activity (Kelly, 1980 and 1983; Kelly, Martin and Pemble, 1984).

The way in which white-collar workers are represented by unions varies. There are some unions that recruit only white-collar workers in a range of companies in one sector of the economy, for example the National Union of Bank Employees (NUBE); there are some that recruit in a variety of occupations across a range of industries and organizations in both the public and private sectors, such as the Association of Scientific Technical and Managerial Staffs (ASTMS); there are those which recruit in one relatively small section of the economy, such as the Inland Revenue Staff Federation (IRSF); and there are those where white-collar workers are recruited into predominantly manual unions, such as the Transport and General

Workers Union (TGWU). The arrangements for the representation of non-manual workers in the United Kingdom again defy easy generalization.

A number of theories have been advanced to explain the differences both between manual and non-manual rates of unionism, and the different rates of unionism among non-manual groups. The most forcible explanations are those which have attempted to link class and unionism. There are a number of such explanations. It has been suggested, first, that middle-class white-collar workers have failed to realize that they have much in common – particularly their employee status – with manual workers. Second, it has been argued that white-collar workers simply want to maintain their social distance from manual workers and their organizations, for example by stressing professional status. Third, others have pointed out that the ethic of individualism is very important to many middle-class people and they therefore see their social and economic advancement as best served by their own individual efforts rather than by collective action. The evidence offers much more support for the second and third views than for the first one (Lockwood, 1958; Bain, Coates and Ellis, 1973).

Differences in the density of trade union membership in different sectors of the economy have been explained with reference to size and character of employing units. It is suggested that white-collar unions prosper where the employing unit is large and highly bureaucratic, or where white-collar workers work in close proximity to unionized manual employees. Similarly it has been proposed that employer and government attitudes are crucial for white-collar unionism to flourish. Finally it should be noted that there is limited evidence that some white-collar workers only turn to trade unionism enthusiastically when their incomes or standards of living are thought to be threatened (Kelly, 1980).

All the above explanations about the nature of non-manual unionism offer insights into why non-manual unions have been unattractive to engineers. White-collar unions are 'class' organizations, in the sense that they represent not just a specialized occupational interest but the interests of non-manual middle-class wage labour as against management and other sections of wage labour. In other words non-manual unions are middle-class organizations in the same way that manual unions are working-class organizations. Many engineers have been unable or unwilling to see themselves as unorganized middle-class wage labour. The hankering after the professional ideal by the engineering institutions further helped to obscure the picture. But this is not to say that engineers are simply blind to the reality of their class position. The research indicates that most engineers, here defined as unorganized wage labour, wish to remain that way because they tend to identify not only with their employers but also with the general goals and aspirations of capital as against labour. The trade unions which have proved attractive to engineers have been precisely those which (publicly at least) have eschewed a class ideology and cultivated a moderate and 'modern' middle-class type of image. A further factor worth noting, for recent years, is that many engineers' individualism and political moderation has made them feel inhibited about joining those unions which operated a closed shop.

The arguments relating to size and bureaucracy and employer attitudes are important. Certainly the largest and most bureaucratic organizations in Britain – those in the public sector – are precisely those where unions catering for engineers have flourished (see tables 10.2 and 10.3 above). However, recent research has suggested that it is not size, bureaucracy or employer attitudes which encourage non-manual unions but specifically employer attitudes to *manual* unions (or to unionism in general). So where managements are favourable to manual unionism among their employees, they are also likely to allow non-manual unions to develop (Prandy, Stewart and Blackburn, 1983). In theory this should favour the unionization of engineers since many engineers do work in locations where highly organized manual workers are also employed. The fact that some organizations have thriving engineers' unions and others do not suggests that employer attitudes alone are not sufficient to explain the absence or presence of unions. There is some evidence to support the view that some professional engineers have turned increasingly to unions as their incomes and standards of living have been perceived as falling. There is some corroborating evidence which shows that where engineers are unionized their salaries tend to be higher than those of non-unionized engineers.

Figures on pay produced by the Engineering Council for the year 1982–3 show that where trade union membership among chartered engineers is high, so is the salary differential between unionized and non-unionized members. This happens to coincide with the fact that trade union membership is higher in the public sector (75 per cent for chartered engineers and 77.6 per cent for technician engineers) than in the private sector (17.9 per cent for chartered engineers and 29.8 per cent for technician engineers). The public sector has a much older tradition of recognizing trade unions for its non-manual employees. Thus where employers conceded union rights and where unions organized substantial numbers of engineers as far as salaries were concerned the results were very good for the engineers. However, the reason for this may have been that public sector pay and conditions of employment were generally better than in the private sector, with public sector engineers just one group of beneficiaries. In any case the situation may be changing in the 1980s.

There is probably no simple answer to the question of why unionized public sector engineers have been relatively well paid. The issues are of the chicken-and-egg variety, with many chickens and many eggs involved. Yet a number of things can be said with some confidence, and they point to some useful general conclusions about the collective organization of British engineers.

First, it is very clear that the position and therefore probably the status of engineers, and of all employees in the public sector, has traditionally been more secure than in the more competitive private one. Another important background factor is the traditionally high status of public service and the lower status of manufacturing. Yet the fact that public sector engineers are more unionized may either be an effect or a cause of this higher security and status, or have little to do with it at all.

Public sector engineers are more likely to belong to professional bodies than private sector ones, partly because their employers have recognized or asked them to join them, partly because of the public sector's traditionally favourable attitude to the professional service ideal. Public sector employer support for the engineering institutions is a further reason why they may be attractive to individual engineers who are worried about their status, in so far as public sector employment and professionalism lend each other prestige. Engineers in the private sector who accept their employers' distaste for white collar trade unions while feeling worried about their own situations are also likely to be vulnerable to the attractions of professionalism. Yet, as we have stated, the professional ideal is of dubious relevance to any engineer who is an employee, whether in business or in the public service, whereas the case for trade unions, although not conclusively proven, does seem a more logical one. Effective trade unions for engineers should of course be real trade unions, not ones which hanker after professional status and 'responsibility' as we have seen UKAPE found to its cost (Dickens, 1972 and 1976).

A particularly good example of a genuine and effective trade union for engineers is the Electrical Power Engineers' Association (EPEA) established in 1913 to represent engineers and other managerial level people in electricity supply. In 1977 the EPEA founded the Engineers' and Managers' Association (EMA) with quiet government encouragement and after several of the engineering institutions had consulted it about a need for effective trade union representation for their members. The CEI had issued a policy statement on professional engineers and trade unions, arguing for the first time that professional engineers should join trade unions, including TUC-affiliated ones. The CEI regarded the EPEA as a suitable union for professional engineers in the private sector and, across all employment sectors, the only appropriate union for them in the TUC.

When EMA was formed it consisted of EPEA and a number of groups of engineers from private sector firms. Today most EMA members still belong to it because of their membership of EPEA but EMA also contains significant numbers of engineers in aerospace, shipbuilding and other sectors, along with former members of the Association of Supervisory and Executive Engineers (ASEE). Of EMA's approximately 45,000 members, just over three-quarters are EPEA-based. Outside electricity supply and shipbuilding EMA has secured recognition agreements from firms in a broad range of engineering sectors. The image of EMA is modern, technocratic and enlightened-cum-responsible. Its General Secretary served on the Finniston Committee and it has continued to be listened to in the areas of engineering education and energy policy (where EPEA has considerable experience) as well as on topics which are more conventionally the concern of trade unions.

Other 'modern', relatively successful trade unions for professional engineers are the Association of Management and Professional Staffs (AMPS), the Technical, Administrative and Supervisory Section (TASS) of the Amalgamated Union of Engineering Workers (AUEW), and the Association of Scientific, Technical and Managerial Staffs (ASTMS). All of these, as well as EMA/EPEA, offer a middle-class type of image to prospective members as well as successful records of collective bargaining over salaries. The list is not exclusive: several other white-collar unions have also done quite well for their professional engineer members. (For more details of managerial-level unionism, with emphasis on the representation of engineers, see Roslender, 1983, and Bamber, 1986. See pp. 3–7 of Bamber for some relevant international comparisons).

Conclusion

The analysis presented in this chapter differs in one significant and crucial respect from most commentaries on the state of engineering as a profession. Conventionally it has been argued that the main reason why engineers have not achieved true professionalism is because employers subvert or pervert the professional aspirations of engineers for autonomy and status in favour of their own needs. It is certainly true that engineers are not fully professional according to the criteria usually used to define a profession, but the central reason lies not in the organizations within which engineers work but within the nature of engineering itself which is not a profession but salaried employment. To assume otherwise, as is conventionally done – even, up to a point, in the Finniston report – is to misunderstand both the historical development of professions and the nature of engineering itself.

A great deal of evidence suggests that most engineers find it natural to identify broadly with the goals of the organization for which they work – profit, product quality, and so on (Child, 1982). Engineers are not on the whole concerned with knowledge for its own sake, or even for autonomy, independence and self-regulation. An engineer's most important peers are those who judge his work in the immediate context with reference to effectiveness and cost. Technical success (a goal of engineers) is reasonably consonant with the commercial and other goals of most employing organizations.

All of the above strongly suggests that professional associations trying to follow the examples of the British Medical Association, the various Royal Colleges of Medical Practice or the Law Society are not perhaps the most suitable vehicles for the collective organization of engineers. However the fact that most engineers are in favour of business aims does not automatically mean that trade unions are unsuitable for them. Many if not all trade unions in Britain are, most of the time, in a relationship of accommodation rather than conflict with employers. Certainly the policies of most trade unions for most of the time since 1919 point to a movement which fundamentally, if somewhat reluctantly, accepts the *raison d'être* of capitalism or free enterprise. Therefore it would seem that trade unions which can offer instrumental benefits to engineers while not threatening the existing social order are likely to serve them effectively in the long term. The professional institutions can continue to play a vital role as study bodies, as sounding boards for engineers' opinions, and as important monitors of and pressure groups for engineering education and training.

While the formation of the Engineering Council has helped to bring more order and clarity to the rather chaotic structure of engineering education and training this will not turn engineering into a profession.

Engineering aims to sell goods in the markets of the world far more often than it aims to provide advice and personal services to individual 'clients'. The creation of a more unified engineering education and qualification system will not change either of these things. But the main point is, surely, that Britain's engineers should be aiming much *higher* than 'mere' professional status. They should be aiming to replace a national elite of arts and natural science graduates which has traditionally been served and more recently been threatened but not replaced by professionals. They should strive to turn themselves into enlightened and tough-minded technocrats, capable of involvement at the highest levels of government and of dominating the boards and managements of companies.

11   Engineering Work: the Division of Labour

Introduction

The central concern of this chapter is with the way in which the concept of the division of labour can be used to explain the position of engineers in British society. We proceed, following Braverman (1974), by describing two sociological meanings of the term division of labour: the social and technical. We then demonstrate the particularly important place that engineers *should* play in the social and economic division of labour by considering evidence relating to social and technical change and development. We next focus on engineers' position in the social division of labour. We point then to the contradictions of many British engineers' roles in the technical division of labour, with particular emphasis on the crucially important function of production. These contradictions, we suggest, are closely related to many of the other problems highlighted in this book.

The Social and Technical Division of Labour

The idea of productive and co-operative activity is central to the concept of the division of labour. A unique characteristic of human beings as distinct from other animal species is the capacity to produce goods and services beyond basic subsistence needs in co-operative relationships. The social division of labour describes the particular ways in which the producers and consumers of different goods and services within a society are differentiated from, and interrelated with, each other. The various forms which the social division of labour take generate the different types of social and political organizations found in a society. The technical division of labour in contrast refers to the specific productive configurations or arrangements in particular employing or work units.

   The existence of society presupposes a social division of labour, since human existence is only possible if material needs are sustained and men and women co-operate with each other to provide and satisfy them. The social division of labour not only describes the differentiation of productive and non-productive functions in a society but also the interrelationships between men and women, young and old, families and institutions and the sacred and profane. Therefore the nature of the social division of labour in any society, at any given time, will be related to the age and sex composition of the family and community; the requirements and demands of the religious culture; the patterns of social interaction and behaviour; *but most importantly*, to the available techniques of production.

The technical division of labour was most famously described by the Scots political economist Adam Smith. He analysed the division of labour in manufacturing. He enunciated the principles of the division and sub-division of functions and tasks in the production process and concluded that it was more efficient for it to be sub-divided into functions and the functions further sub-divided into tasks and for each operative to do one repetitive and simple task, than for one worker to make a whole product.

Of the nail-making industry, Smith observed:

> In the way in which this business is now organised, not only the whole work is a peculiar trade, but it is divided into a number of branches, of which the greater part are likewise peculiar trades. One man draws out the wire, another straights it, a third cuts it, a fourth points it, a fifth grounds it at the top for receiving the head; to make the head requires two or three distinct operations; to put it on is a peculiar business, to whiten the pins is another; it is even a trade by itself to put them into the paper; and the important business of making a pin is, in this manner divided into about eighteen distinct operations.
>
> Adam Smith, *The Wealth of Nations*, pp. 4–5

The technical division of labour is influenced strongly by the available techniques of production and knowledge of materials used in the production process. These in turn affect the length of the chain of command, the span of managerial control, relationships between 'advisers' and 'doers' in management, ratios of managers and administrators to total employees, the proportion of manual to non-manual and skilled to semi- and unskilled workers in a productive organization. In other words the social system of a factory or any other productive unit is determined by the nature of its technical division of labour. Engineering should play a crucial not to say pivotal role in both the social and technical divisions of labour by virtue of its concern with the techniques of subsistence and manufacture. However in practice different societies and cultures ascribe somewhat different roles to engineers and engineering. In the next section of this chapter we argue the case that engineers *should* play a crucial role. In the succeeding sections we discuss how, in Britain at least, this central role is partly denied. (For a full discussion of the concept of the division of labour see Esland and Salaman, 1978; Giddens and Mackenzie, 1982; and Lenski and Lenski, 1978.)

The Role of Engineering in Social and Technical Change

In order to demonstrate and substantiate the view that engineers do play a crucial role in the social and technical divisions of labour we now consider some arguments about the role of engineering in social and technical change.

The constantly changing and evolving nature of human societies has been widely noted and various theories have been advanced to explain the process. The view of social and technical change to which we subscribe is that the development and evolution of human societies occurs as a consequence of the interaction between men, the physical and social environment and man's ideas and beliefs. This process is therefore almost infinitely variable and this accounts for the uneven and different rates of development among different societies (Giddens and Held, 1982; Lenski and Lenski, 1978). We do not believe that social and economic change has a single cause or that it is uni-directional. However, there are some writers who do, and before proceeding with our argument we must dispose of at least one particularly virulent strain of mono-causal explanation. Specifically we need to dispel the myth that science is *the* motor of social change and development.

The model offered of the pre-eminence of science in the developmental process is as follows. The scientific researcher working in an academic situation makes discoveries and generates general scientific data, principles, theories or laws. These are, in their turn, translated into something called technology which in turn produces new production processes or new types of hardware. Knowledge is divided into pure and applied strains, with the pure being paramount because the applied depends upon it. New scientific knowledge, in this view, is seen as the source of technical change, which in its turn is the source of social change. Technical change is seen as being relatively sudden, sometimes revolutionary and as being dependent on new scientific discoveries. Central to this line of reasoning is the assumption that engineering is the application of science.

The argument that science leads to technology, which leads to new forms of hardware, which in turn lead to social change, can be challenged on two main fronts. First, there is little historical evidence that the process actually occurs in this way and, second, the notion that science is about the discovery of universal laws or the generation of theories is highly questionable (Kuhn, 1962). The view we wish to argue for emphasizes the role of *homo faber* – man the creator – not man the thinker, and also recognizes the interaction between man and his social and physical environment (see Mumford, 1966; Pavitt, 1979 and 1983; Rolt, 1970; Sorge and Hartman, 1980; Rothwell, 1981).

Every culture, be it pre-industrial or industrial, has a substantial store of information about the environment to which the society must adapt. The information is about coping with life and defining life's problems and is not infrequently imbued with religious significance. This is not primarily scientific knowledge, although it does often include it. Rather it is practical or technical knowledge which serves, more or less efficiently, the purposes and needs of a community. In every society a large component of this information concerns food, where and how to get it and how to process it for consumption or storage. There is also knowledge about other material resources available to the society and how these may be converted into useful forms such as fuel, clothing, shelter, tools, weapons and ornaments. All of this information is technical knowledge. It concerns the use of material resources to satisfy human needs and wants. This type of knowledge has been far more important than any other type in shaping development and change. Technical advance is partly, sometimes crucially, responsible for population growth, the creation of new symbol systems, the growth of vocabularies and systems of ideas, the increased production of material goods and the complexity of social structures. The techniques available determine the range of what is possible, and the relative costs of the options open to a society. Normally the process has relatively little directly to do with science or the use of scientific principles, but is instead associated with creative ingenuity or *Technik*.

Engineering (the etymological root of which means ingenuity) is about the art of making things that are useful and work. The outputs of engineering are three-dimensional, concrete and real. The criterion by which they are judged is usefulness. Engineers are more concerned about whether things work, rather than why they work. In this, the *Technik* view of technical change, which opposes the previous 'Science leads to Technology leads to Hardware' (S → T → H) one, technical change is viewed as a response to socio-economic and technical problems and it consists of gradual improvement (Sorge and Hartmann, 1980). Human society learns by doing and technical ideas and procedures are spread by the processes of continual absorption and modification. Scientific knowledge is very often used but normally only in an ambient, background sense. In many instances, however, the changes are made without new scientific knowledge, because it does not exist – there is no science to 'apply'. The fact that the piece of hardware or machine works is enough; it doesn't really matter if we do not know why it works. Most technical change is in fact undramatic, drawing on previous technical change in an incremental, piecemeal fashion (cf. Jewkes, Sawers and Stillerman, 1969; Langrish *et al.*, 1972).

Technical changes occur mainly because of an effort to save money and/or effort (in the short-term) and as part of a more general desire to increase human skills and comfort.

The engineer's and craftsman's role in this process is critical and central. It is their function to bring together extant products and processes and it is from them that newer configurations of existing processes emerge. It is their job to harness existing technical expertise to new economic and social problems and to demand identified in markets and it is their function to ensure relative ease, quality and reliability of manufacture. In theory this case for the engineer's importance is undeniable. When considering data about the general position of engineers, however, a rather different picture emerges and it is to this to which we now turn.

Engineers in the Social Division of Labour

All groups and individuals are assigned a place in the social division of labour. However engineers have, in principle, a pivotal place given their involvement with the techniques of subsistence and survival. In terms of conventional sociological, historical and economic scholarship, there are two rather different approaches to describing and explaining the dual issues of the centrality of the engineering dimension in a society, and the differential status, income and other rewards which accrue to engineers in different societies. One, the prescriptive school, tends to focus attention on some particular quality or trait of the occupation of engineering both to highlight the engineers' central role and to try to explain his cultural status. The other, the explanatory school, concentrates on the broader issue of the class position of engineers. We consider each approach in turn.

The prescriptive school can be subdivided into what we call technocratic, professional and managerial versions. Each concentrates on the aspect of engineering regarded as of most importance by the commentators. The technocratic view states that engineers are best described as broadly educated and technically specialized top-job holders who have reached an elite position culturally, socially, economically and politically by virtue of expertise. Technocrats are thus defined as those whose possession of high levels of education and skill gives them a central and dominant role in the social, economic and political system. The decisions of the technocrat are said to have political consequences far beyond the immediate workplace. Technocratic theories were originally developed in societies like France where engineers enjoy very high status, prestige and power.

The professional version developed in societies where the importance of the professional ideal is a dominant one. The argument is that engineering *is* a profession by virtue of the specialized knowledge which engineers have and that the 'applied' skills of an engineer are analogous to those of other learned professions. This version also purports to describe the centrality of the engineer's role in society. However in spite of engineers' specialist knowledge and skills most do not enjoy the status, prestige and income comparable to that of doctors, lawyers and accountants. The reason for this, it is asserted, is that engineers are not professional enough. Reforms of engineers' professional bodies, changes in educational institutions and syllabuses are the typical prescriptions (cf. Finniston, 1980). Not surprisingly, this view is most commonly stated by Anglo-Saxon commentators in whose societies professions thrive and engineers often have a relatively lowly status.

The final prescriptive approach links engineering to management. The view is premised on the notion that in modern societies managers play an important strategic economic role. This role has developed partly because of the financial structure of modern business. Because ownership is vested in large numbers of shareholders it seems to mean that the members of the group which administers enterprises on behalf of the shareholders are the *de facto* holders of industrial and commercial power. It is further argued that engineers are a significant element in management, on two counts, first the fact that engineering by definition involves managing people and things, and second by virtue of the importance of technical efficiency in a productive unit. These views tend to be propounded by admirers of American, Japanese and West German industry, although management is so engineer-dominated in Japan and West Germany that it somehow seems that it barely exists in its own right there.

Whatever the superficial attractions of the three views just outlined as a means of analysing the position of engineers in the social division of labour, each is inadequate in part. The inadequacies stem from internal problems with the ideas themselves and the analytical status of the explanations offered.

In each prescription there are internal inconsistencies which detract from their descriptive accuracy. In the technocratic view the assumption is made that knowledge and skill give power. How knowledge may or may not give power, and the fact that financial rather than technocratic elites still very often control industry, or that where technocratic elites do hold sway it is because the powerful have become knowledgeable rather than the other way round, are problems which are conveniently ignored. As far as the professional view is concerned the lack of precision in the use of the term profession coupled with a culture-based view of the importance of professions effectively confines the analysis to English-speaking countries. The assumption underpinning the very influential managerial thesis – that a new stratum of strategically crucial managers has emerged to replace or envelop financially dominant groups and owners across most economic sectors has never been empirically demonstrated. The details of the managerial view may be analysed as follows.

The development of a large-scale 'managerial stratum' of technical and administrative experts standing between capital and labour, serving one and superintending the other, is mainly a phenomenon of industrialization and the last century or so. This suggests that it is useful to think of a historical development whereby growing numbers of expert managerial specialists have been absorbed into ever-larger employing units, without very seriously altering power structures in the long run or career structures in the short, or the assumptions about authority which underpin both.

As far as larger British manufacturing units are concerned this pro- cess of absorption might usefully but very tentatively be portrayed as having consisted of four main stages. The first, the owner–manage- ment stage, was the norm before the First World War, when most 'management' was performed by owner–managers aided by internal sub-contractors and/or foremen, and helped and advised by relatively small numbers of clerical and technical people. In the second, 'line- staff' stage, between the First and the Second World War, 'staff' functions such as accounting, sales, design, technical development and welfare were increasingly organized into semi-autonomous de- partments to provide services for the main dominant 'line' functions of production and sales. Financial tasks are increasingly regarded as important.

The third, generalist–specialist stage which lasted for perhaps two decades after 1945, witnessed growing attempts to recruit and groom individuals as potential top job holders. Functional departments were larger than in the past and increasingly staffed with qualified people. The status of the 'line' functions of production and sales was in decline, but marketing and financial expertise were very highly regarded. The fourth, final and present stage, the 'defensive' one, seems to have begun in the mid-1960s. Here public concern about manufacturing performance foreshadowed a slow down and perhaps a reversal of the decline of the status of production and sales. The growth of the managerial stratum continued but in a more self--conscious and increasingly self-critical way. There was considerable uncertainty, indeed insecurity, concerning the legitimacy of authority and pay differentials, loosely associated with a (probably temporary) rise in the status of personnel/industrial-relations work. There was growing concern about the calibre, potential and attitudes of job holders, especially in production, technical development and sales (Glover, 1977a; 1979).

This picture is elaborated in part of the work of Littler and Salaman (Littler, 1982; Littler and Salaman, 1984). British manufacturing firms through most of the nineteenth century were typically (not entirely) small, better understood of as workshops rather than as factories, with both capital and labour employed on a small scale by present standards. Many of the larger nineteenth-century British employers were not the immediate employers of 'their' workers; rather they used the 'intermediate internal contractor who had a contractual relationship with the over-arching employer, and in turn was an employer of labour himself' (Littler, 1982: 65). Employers provided raw materials, the fixed capital and most working capital, and controlled sales. Contractors took workers on, supervised them and laid them off, and were paid lump sums by employers for work done, keeping some money for themselves after paying wages and (often) working capital. Elsewhere, owner–managers employed and supervised workers with the help of foremen and, by later standards, small numbers of clerical and technical staff. Between the two World Wars, the bureaucratization of larger British industrial firms took place in a period largely characterized by depression and with accountants and other efficiency experts (defined broadly) in the ascendancy (Armstrong, 1984b).

The idea that the above developments could be described as ones in which expert, professional, salaried managers have taken over the reins of power from yesterday's supposedly rapacious owners of dark Satanic mills has been argued systematically since the 1920s. It is both plausible and popular, given this century's growth of large-scale organizations and its improved living standards and working conditions. It is also true that ownership of company shares in particular and of the means of production in general is more widely distributed than it was in the past.

However, the idea of a managerial revolution whereby private ownership of the means of production is no longer the root of all evil has been depicted as being misleading for two major reasons. First, several British and American studies have shown that owners and managers do not have very different aims, both tending to seek short-term corporate profits and growth and/or a relatively quiet life. Both also believe in the long-term maximization of profit, partly because profit figures are on the whole very useful indicators of performance, and partly because making money is regarded as desirable for all the parties involved. Such views appear to be popular amongst most kinds of businessmen or managers irrespective of what they own personally. Also, the more senior the manager, the greater the tendency to·believe, ostensibly at least, in a fairly liberal and 'enlightened' way, in the established system of power and authority (Nicholls, 1969; Stanworth and Giddens, 1974; Zeitlin, 1974; Nyman and Silberston, 1978; Scott, 1979 and 1982; Fidler, 1981). Thus the fact that corporations have become much larger and more bureaucratic than in the past does not generally seem to mean that the managerial and technical tail nowadays wags the strategic dog of the top executives and owners. The constraints of the market and of the system of ownership force managers to pursue profits whether they want to or not. Managers are cogs in the bureaucratic machine; they 'control' labour but they are themselves a superior form of labour, controlled from above within their units and from outside them by the system.

However in the final analysis it is the criticisms of all three of the prescriptive theories (technocratic, professional and managerial) at the *analytical* level which are the most telling. First, in their various ways, each prescription is culture-bound. The particular trait chosen to describe engineering is not particularly useful for describing engineers outside the society where the theory was developed. In fact the theories often tell us more about the cultures in which they were developed than about engineers, since as noted technocracy is largely a Continental thing; professions and management are largely Anglo-Saxon ideas. Second, all three tend to concentrate on the non-engineering side of the engineering task – theoretical knowledge in the case of the technocratic and professional views, and man– and thing-organization in the case of the managerial view. The crucial feature of understanding the pivotal place of engineers in the social division of labour lies in the engineering dimension itself – the making and utilization of useful artefacts. In all three prescriptions, this central point tends to be sidelined. Finally the three views also tend to be descriptions and not very good descriptions at that – they explain little because a mixture of description and prescription is substituted for serious analysis (see Child *et al.*, 1983; Giddens, 1973; Kelly, 1980; McQuillan, 1978; Scott, 1979; Thackray, 1981b; Thust, 1980).

A much more fruitful way of examining the place of engineers in the social division of labour lies in an explanation of the class position of engineers. In order to understand this a brief examination into the meaning of the concept of class is required. Class systems emerge out of the division of labour in the sense that class relationships are based on economic domination and subordination, exploitation and emiseration which occur in the co-operative activity of production. Classes originate in surplus production, the production of goods and services over and above the basic subsistence needs of the population. In the earliest and simplest types of society there is little surplus production. (What animals are caught are killed and consumed; surplus is stored in other men's stomachs.) In such societies there is no private property. Production is based in these simple societies on a very elementary social division of labour and on the communal sharing of property. The fruits of productive activity are distributed through society as a whole. Class divisions arise when a surplus is generated and products are exchanged in markets and it becomes possible for a class of non-producers to live off, either wholly or in part, the producer activities of others. Eventually those who are able to gain control of the technical bases of producing the surplus will form a dominant or ruling class. Once dominance is established the ruling group must solve the problem of maintaining their dominant position. They may do this through coercion, physical threats, persuasion or some other means (Giddens, 1973; Giddens and Held, 1982; Lenski and Lenski, 1978).

Three elements are critical in understanding the nature of class in a society – production, consumption and the extent to which the class system is rigid or fixed. Class is grounded in the economic relationships which are generated by the means of production, distribution and exchange in the social division of labour. Classes may represent consumption groups and patterns of living. Wedded to both the productive and consumption dimension will be a greater or lesser awareness of the degree to which people are members of classes. Additionally this awareness may sometimes, though not invariably, lead to a willingness to engage in political action. In any society the class system will be more or less rigid or open. The extent to which a class system is open and flexible will depend upon the degree to which the dominant group excludes others from membership and the opportunities for mobility between groups (Parkin, 1974).

It is well beyond the scope of this book to enter into the highly technical and sophisticated debates about contemporary class systems. Below we only sketch the major features in the development of the class system of Britain as a means of highlighting the particular position of engineers in that system. The nature of the class system in

all industrial societies is a product of the transformation from a predominantly agricultural–rural society with a relatively simple
division of labour to predominantly industrial and commercial–
urban societies with a very complex division of labour. In all societies
that have undergone such a transformation, the groups conventionally described as the middle classes have both increased in number,
and become more highly differentiated, while in Britain but not
always elsewhere the upper classes have tended both to retain their
dominant political and economic positions and maintain their social
exclusiveness. Schematically what has happened is that the groups
who used to occupy the middle ground (the farmers, shopkeepers,
merchants, the independent professions of law and medicine, the
military, the clergy and the master craftsmen) between the very rich
and the labouring classes have been numerically overwhelmed by
newer groups aspiring to and achieving middle-class status. In the
eighteenth and nineteenth century a class of entrepreneurs and
businessmen and salaried (as opposed to independent) professionals
appeared upon the scene. Later in the nineteenth century and
throughout the twentieth another much larger group of individuals
aspiring to middle-class status arrived, namely the vast number of
clerks and office workers and formally qualified experts in an era of
much greater levels of literacy who were and are mainly employed in
the public and private bureaucracies which have flourished particularly in the twentieth century. From the second quarter of the
twentieth century onwards the ranks of the middle classes have been
swollen again by the technical and scientific personnel employed in
public and private organizations and by various new or growing professional and semi-professional occupations – teaching and social
work for example (Dahrendorf, 1959; Bain, 1970).

   The lower stations of pre-industrial society – the mass of labouring,
agricultural, poor and the small number of skilled tradesmen, have
been transformed into a mainly urban working class distinguished
into three distinct groups: a highly skilled group enjoying stable
employment, job security and a relatively enhanced income and
position; a larger group of semi-skilled and unskilled workers who
also enjoy relatively advantaged wages but are much more prone to
the fluctuations in national and international economies; and a low-
skill group with low pay, little job security and frequent unemployment. While the last of the three groups is considerably disadvantaged
vis-à-vis the other two, all three share the characteristic of being in
a subordinate position at work and at the end of the managerial
chain of command (Edwards, 1979; Wood, 1982).

   Qualified British engineers are not part of the upper-class elite as a group, although individual engineers may have been born in such a group. Their technical skills and the nature of their tasks do not inevitably make them an economic, political or socially dominant group. Neither are engineers part of the working class. They are not in a totally subordinate position, nor generally speaking do they have the same market disadvantages described above. Engineering as an occupation is mainly middle-class, even in countries in which engineers are very powerful. Engineering as a distinct occupational category could be thought of as historically part of the master-craftsman group and contemporarily as the dominant group of technical occupations. In terms of salary and prestige engineering is clearly middle or upper-middle ranking. Engineers as a group often lack the political and economic power of the old middle classes but enjoy considerably enhanced chances in life compared with the working classes. It is true, of course, that the power and prestige of engineers varies between societies, but the essentials of the class identity and hence the position in the social division of labour remain across cultural boundaries. The engineer is strategically placed in the social division of labour by virtue of his skills in dealing with the techniques of subsistence. But even in countries where engineers form the most prestigious occupation of all, the majority are generally servants rather than wielders of power Their work may transform society but only a minority pull the strings of political power. At the same time the engineer does often dominate others in the activities of transforming things, in making artefacts, and in constructing machines that work. We conclude therefore that engineering plays a pivotal role in the social division of labour – but that the political differences in the power accorded to engineers vary according to cultural conditions (Child *et al.*, 1983; Hampden-Turner, 1983; Lawrence, 1980; Sorge and Warner, 1986; Whalley, 1986).

Engineers in the Technical Division of Labour

In the previous section reference was made to the fact that the work of certain commentators whom we referred to as the 'prescriptive' school had suggested the existence of various problems for engineers. We have rejected the solutions offered by these writers, arguing that to understand the position of engineers in the social division of labour, the best way to proceed is to consider their class position. Notwithstanding our earlier criticisms, these writers have nevertheless highlighted an important issue, namely the diversity of rewards which accrue to engineers in different settings. This is the issue to which we now turn. However, rather than seeking the answer to the question by considering the role of engineers in the social division of labour, a much more fruitful way of contemplating the issue is by considering the engineer's role in the technical division of labour. The argument we now advance may be schematically outlined. First we assume that, given the centrality of the engineering dimension, engineers should play an important role in the technical division of labour. Second, in some societies (notably Britain) certain factors prevent engineers doing this. Third, the reason for this is to be located in cultural and social phenomena which can be easily identified (though not easily changed).

This argument is important because we are highlighting a basic disjunction between the needs of a society and, in the British case, the technical division of labour which aims to produce those needs. In essence we argue that in Britain engineers become, partly by virtue of the organizations they work in, predominantly concerned with activities which are peripheral to the engineering dimension.

To build our argument we go first to what is often regarded as the hub of the engineering dimension, to production. Productive activity is crucial to a society's continued existence. It is the central activity in manufacturing logically and chronologically (Lawrence, 1980 and 1986), between the 'push' of design and development and the 'pull' of the market. Further, we have stated that the techniques of satisfying the relevant needs should primarily be the province of the engineer. In an advanced industrial, indeed any, society production is therefore at the very hub of human work. We now consider the evidence concerning production in the United Kingdom compared with some other countries.

Production is concerned with activities which must be accomplished so that a finished artefact can come into existence. The technical division of labour involves two aspects: fabrication, the manufacturing activity undertaken by operatives with the help of machines, and the directing and supervising activity undertaken by managers (including foremen). The directing activity includes such things as the location of facilities, organizing layout and materials handling, deciding on and checking work methods, scheduling operations, managing inventories, supplying resources, quality control and measuring and evaluating performance.

As just emphasized, production is the central function in manufacturing industry because commodities must not only be designed and developed but produced before being sold and distributed. Production provides the tangible and measurable output of a company; most of the plant, machinery and equipment is in the factory for the purpose of production and most of those employed in the company are engaged directly or indirectly in productive activity. We know that many engineers argue that design is more central to their work, and that some social scientists argue that 'research and development' (including design) is. But engineers are only right here in the sense that design is *their* core skill. The 'research and development' argument contains the same kind of a grain of truth; but it is based on the dubious Science → Technology → Hardware model of technical change (rather than manufacturing as a whole) and it manifests a rather snobbish disregard for details of events at the sharp end. Thus it seems to assume that the clever chaps lay down the guidelines, the 'principles' of manufacture, the methods to use, and then less clever chaps 'merely' design in the details, and make and sell the products.

Production is not unusual in that it is simple in concept but often extremely complicated in execution. In theory it consists of getting materials in, making something with them, and getting the product out of the factory door. In practice the processes involved are normally very complex and difficult, often inordinately so. In practice too, supplies and parts are often incomplete, damaged or otherwise defective, late, and/or completely different from what was wanted. Subcontractors often prove very unreliable. Goods inwards inspectors often seem very slow. Manufacturing frequently involves modifications to suit particular customers. In some industries most products are subject to continual modification. Some designers habitually interfere with manufacturing by changing drawings and specifications at the very last minute – or even after – all with the best intentions. Production planners, production engineers, and industrial engineers or work study staff are always around to interfere with schedules, methods and processes, and with rates of production and remuneration. Moreover none of these are people problems as such, yet that is what production is probably most notorious for, in Britain anyway.

One such problem is maintenance, commonly undermanned, short of funds, mainly corrective rather than preventive, and nevertheless with a superiority complex because it is staffed by time-served craftsmen rather than by semi-skilled production operatives, and often under the thumb of a technical director rather than a production one. Other people problems include absenteeism and sickness, skill shortages, disagreements about overtime working, and all other possible kinds of 'industrial' relations and cover and manning problem. Production managers' dealings with finance, marketing, personnel, quality control and management services departments are often conflict-ridden and bureaucratic. They are continually told that budgets are overspent, or that they are making too many of one thing or too few of another, or in breach of safety regulations, or turning out too many faulty items, or not using machinery properly (Lawrence and Lee, 1984).

Most studies of managers' jobs show that a lot of their time is spent talking: this is dramatically true of production managers who some-times seem to spend almost 100 per cent of their time so doing. This is no doubt why many senior production managers, who may have to write fairly long reports every so often, often do them at home after work where they can find some peace. The dependence of production on other functions is usually very great, and their most obvious conflicts seem to occur with staff in maintenance, research, design and development, and sales and marketing, although relationships with production controllers and purchasing can involve a lot of friction too. Relationships with finance and personnel seem not to be so obviously fraught in spite of evidence that it is their functions which have, over the years, been most strongly associated with decline in the status and power of production. The reasons are probably that they *are* so powerful *vis-à-vis* production, which has grown used to the situation, and also that their day-to-day involvement with produc-tion is in any case nothing like so great as that of the other groups.

In Britain production tends to be separate from other kinds of engineering. Most studies of those engineers who work outside production, in maintenance, design and research and develop-ment have been concerned with their attitudes to work rather than with what they do. This is unfortunate because their work is obviously very important in itself. Even when work content has been studied, writers have too readily assumed that the 'glamour' roles of invention and innovation are more significant than the more usual tasks of design and general technical development. Researchers have often studied such people as staff specialists (perhaps because re-searchers like studying people like themselves) rather than as full, bona fide members of the management team, although they form a large and important component of it. Although most are specialists of one kind or another, so too are virtually all other managers. Indeed about three in every five members of samples of Britain's professional engineers regularly used to define themselves as managers rather than specialists (Glover, 1973 and 1979; see also Bamber and Glover, 1975). This is particularly important, insofar as it suggests that the classical English-language means used to examine the topics covered in this book are often defective. Thus 'managing' is assumed – but not by engineers according to this evidence – to be different from 'engineer-ing'.

Because of the particular constraints on the British production manager it means that he tends to excel at certain activities and has a thorough understanding of the company he works for. The British production manager tends to be a resourceful fixer with a capacity for creative adaptivity to circumstances in spite of adversity. The particular skill the production manager must possess is the ability to deal effectively with shortages of materials, components, packaging, space, time, and broken-down machinery, only the latter being an engineering task (Lawrence, 1980).

While these features may be praiseworthy in themselves the data suggest a number of severe difficulties too. Many commentators have reported that senior production managers often have a large number of different and sometimes conflicting sub-functions under their control. Furthermore it appears that few companies treat this as a serious problem and that the production tends to be neglected by senior company management which views production only as a cost to be minimized. In many British companies, senior management has shown little desire to become involved in the details of design, production or the quality of products. Another consistent finding is that production managers tend to be dogged by conflicts with sales and marketing on the one hand and dominated by the financial function on the other. It has also been demonstrated that most production managers have little or no control over their own production budgets and nor do they have very much influence over or involvement in production or marketing policy or strategy. Few British companies appear to merit Lawrence's (1980) description of German ones, as 'production (all technical ones) plus sales and finance', with the latter two areas far more engineer-influenced than is the case in Britain (cf. Cunningham and Turnbull, 1981, on a relative lack of technical expertise in sales and marketing in Britain).

Many senior executives see the production supervisor embedded firmly at the end of the executive chain. He is expected to secure the maximum shop-floor effort to reach improved targets. He is assumed to mediate the policies of higher management in face-to-face contact with the shopfloor and conversely to feed upwards the information that higher management needs to know for policy making. Along the industrial relations interface the manager is expected to defuse potentially dangerous issues. This task is made the more difficult because most production supervisors and managers do not have much managerial authority. In policy terms they are frequently bypassed by senior management and rarely consulted in decisions which affect their jobs or departments.

Another consistent finding is that there is often little mobility in or out of production. Recruitment into production has tended to be on the basis of knowledge and experience rather than qualifications and there is often a distinct lack of enthusiasm by employers for recruiting graduates into this area. Senior production managers are much less likely than their colleagues in marketing and finance to have high levels of qualification while at first-line level the majority of production managers have no formal educational qualifications (Child and Partridge, 1982). Many studies have reported that it is virtually impossible to recruit academic high-fliers into production. As regards training there is some evidence to suggest that many employers regard the training of all staff, but production staff in particular, as a luxury rather than a necessity. Indeed there is evidence that the amount of in-house training received by production staff has decreased over the last twenty-five years (Foy, 1978). Specifically in terms of management education, production management and manufacturing are the least taught subjects in the Business Schools and Management and Business Studies departments at undergraduate or postgraduate level. Moreover what *is* taught is to some extent irrelevant, concentrating as it does on elaborate numerical prescriptive techniques rather than on the firefighting and political skills and above all technical detail most needed in production management.

Finally and perhaps most damning of all it has been repeatedly demonstrated that in terms of performance as measured by the timely meeting of delivery dates, accurate planning and scheduling, introducing innovation and overall co-ordination of the productive process, the British production manager compares unfavourably with his European and Japanese counterpart. The result is that the typical British manufacturing unit seems to be characterized by large batches of finished or semi-finished products, with components lying idle on the shopfloor and the concomitant tying up and waste of capital.

The production manager is in a dilemma which arises out of his inappropriate position in the technical division of labour. There are a number of conflicting and different pressures on him and at the same time he has neither the personal nor the positional authority to execute his duties effectively. The pressures are associated with work schedules, timing and getting the job done. Sometimes he may have to meet a number of different demands simultaneously while fending off other demands from personnel, marketing and finance. Top management does not delegate sufficient power or authority to the production manager. From below two forces are at work to counter the production manager's aims – on the one hand more or less formal systems of counter-control exerted through the trade union and

primary work groups; and on the other informal means of circum-
venting managerial controls which develop in all work situations
(Roy, 1969). While the production manager is usually well aware of
both the formal and informal constraints on his authority he is in effect
powerless to do much about them. The production manager's
authority is not legitimated either by his superior or by his subordinates (for a discussion of these issues see British Institute of
Management, 1976; Business Graduates Association, 1977; Child,
1978; Child and Partridge, 1982; Foy, 1978; Gill and Lockyer, 1979;
Hayes, 1981; Jenkins, 1978; Lawrence, 1980; McQuillan, 1978; New,
1976; Prabhu and Russell, 1978 and 1982; Thust, 1980; Sorge, 1985;
Sorge and Warner, 1986).

The question arises, then, if engineers are not doing production and
if they are not involved in making things, what are they doing? There
are two answers – first a good proportion are engaged in design and
technical development activity, and second the others are engaged in
some kind of management activity.

We consider first the case of engineers doing design and technical
development. Some 23 per cent of engineers classify themselves as
engaged in research, design and development (Engineering Council,
1983), with nearly half of these in design. A problem is that a lot of
activity is at least partly irrelevant to most of manufacturing industry.
Britain's research and development (R and D) efforts are highly
skewed towards certain sectors like aerospace, defence (especially
electronics), civil aviation and nuclear energy. There is very little
evidence to show that Britain's preoccupation with R and D in these
areas has produced many corresponding national economic benefits.
Indeed small firms outwith the mainstream funding of R and D have
in fact accounted for a far greater proportion of innovation than
those which attract the lion's share of R and D funds. Such expenditure is not correlated with economic growth in Britain. The areas
where the bulk of R and D activity is concentrated in Britain account
for only a small part of Britain's exports (Pavitt, 1979 and 1983;
Rothwell, 1981). Yet a disproportionate amount of talent (including
engineering talent) is located in these sectors. Reports have shown
that many British engineers seem to prefer a relatively uncomplicated
life in R and D and for many years one of the most common
routes into employment in industry for engineering and science
graduates has been through it. Yet a controlled 'laboratory' environment is hardly the best place to learn about risks, markets and
the vicissitudes of production management, although these
should be seen as highly relevant to all forms of technical innovation.

There are relatively few empirical studies of the work of British engineers, in production or elsewhere. However the data which are available present a reasonably consistent picture showing many engineers involved in managerial tasks. The studies show that such engineers spend much of their time in man-management and dealing with human as well as technical problems. After initial employment in a technical function promoted engineers typically tend to be moved to the middle levels of organizations, and there to be mainly pre-occupied with current or specific issues rather than with strategic or long-term planning. For many engineers the key parts of their jobs are related to man-management, inspection and coping with crises which makes job content highly fragmented rather than concerned with strictly defined engineering matters. The data further reveal that many engineers see their work as concerned with trouble-shooting and disturbance handling and that it is characterized by brevity, fragmentation, industrial relations problems, vexed relationships with the other functional areas; and the need for responsive and instant decision-making. For engineers in more structured areas like main-tenance, design, construction and technical development the picture is often one of a more ordered and predictable working environment. However within all the functional areas the various studies show that the engineer's work is dogged by communication problems. This does not mean that engineers are any worse than anyone else at com-municating, an idea suggested, in our opinion quite wrongly, in some recent studies of engineers. However because of their contradictory position in the division of labour they may well experience problems which make communicating difficult.

The reason why engineers find themselves with such problems lies in the British pattern of organization in both the manufacturing and non-manufacturing sectors. This is characterized by formalized pro-cedures which invest such functional areas as finance, personnel and even marketing and sales (but not production) with professional and even managerial authority over engineers. Impersonal principles of hierarchy and irrelevant ideas about professional jurisdiction are appealed to, in order to justify the right to manage by non-engineers. Many British managers seem to be more concerned with authority derived from their so-called professional expertise rather than with the authority of their roles and tasks. In short the emphasis on profes-sional autonomy and authority may result in the displacement of effort from the central business of making and selling things.

Thus British companies tend to be characterized by a high degree of occupational specialization in the non-productive functions. The specialization is aided and abetted by the ideology of professionalism and an education system which has traditionally devalued engineering and production. The result is that many British managers show less concern for the needs and standards of production than for their professional specialism. Some British companies carry extremely high administrative overheads in the form of personnel not directly engaged in the production process. In those companies which employ a large number of professionally qualified administrative staff and where therefore the relationship of direct labour to indirect labour is balanced in favour of indirect labour, the more standard rules, procedures and paperwork tend to be generated. In effect many such rules and procedures amount to nothing more than externally imposed controls on production. Management becomes the art of surviving in a world of ritualistic rule-following backed up by appeals to spurious professional expertise (Child, 1978; Child and Partridge, 1982; Child *et al.*, 1983; Kotter, 1982; Limprecht and Hayes, 1982; Pavitt, 1980; Thust, 1980; Sorge and Warner, 1986).

The organizational forms and the attitudes of job holders in organizations do not exist in a vacuum. They are also a product of the attributes of the job holders within them and the wider society in which they operate. To substantiate this argument we refer to the managerial task and the social and educational backgrounds of the people who carry out that task.

In general terms, and after the maintenance of the existence of the employing unit, management is concerned with directing and co-ordinating actions. As such, managers exercise power and authority over subordinates who more or less accept the directions given. Managerial authority cannot be taken for granted, in the long run, and so managers have in effect to find strategies to hold on to their authority. Aside from direct coercion or physical threats, one strategy is to be seen to be able to do the directing and co-ordination efficiently. This has been a German and Japanese solution. Another strategy has been to appeal to management science or professionalism. This has by-and-large been a British and American solution. The management science approach in which the 'science' revealed certain prescriptions as to what management ought to do is now all but discredited by the studies of managerial work. These studies of what managers actually do show that, far from being a decision-making planner who organizes his staff to rationally maximize outcome, the typical manager was in effect a responsive, flexible manipulator of people, information and things. The definition of management to which we subscribe is that the manager is a focal point in his unit whose major skills include elements of trouble-shooting and firefighting. Although it does of course, include a great deal of planning, organization, decision-making and so on, management is nevertheless an art or craft rather than a science and much of it is about doing the simple things properly. Management is also about coping with uncertainty amidst a great variety of potentially relevant information and about getting things done through a large and diverse set of people despite often having little or no direct control over them. The effective manager learns most of his job on the job. It is the only effective way of mastering the particular skills required for particular organizational contexts. This implies that effective managerial skills are sector-specific skills. In Britain our education process produces not sector-specific type skills but rather different animals (Hayes, 1981; Kotter, 1982; Lawrence, 1980; Limprecht and Hayes, 1982; Miners, 1982; Mintzberg, 1973; Thackray, 1978, 1981a, 1981b; Thickett, 1971; Thurley, 1982).

This is revealed when data on the social backgrounds of managers are inspected. Notwithstanding problems of sampling, classification and research design, with the extant data on managers, a number of generalizations about their backgrounds, educational and social, can be made. As far as senior management in the United Kingdom is concerned the social origins (as measured by father's occupation) are not particularly upper-class, with the majority having fathers in middle-class, non-manual occupations. Senior management in the private sector tends to be staffed by men and women from well-to-do backgrounds rather more than the public sector does and the financial sectors and function are similarly more elitist than industry and commerce as a whole. Middle managers tend to be more humble in terms of social origins than their senior colleagues with a majority having fathers from the manual working class. At the lowest levels of management (first-line and supervisory) three-quarters have backgrounds in the manual working class.

In general terms top British management tends to be under- and less relevantly educated compared with similar job holders in mainland Europe. The majority of British senior managers have attended local education authority grammar schools or equivalent with a sizeable minority having attended independent schools. Those educated in the private sector of education are the ones who in statistical terms are the ones most likely to reach board level. Nearly 12 per cent had studied science, 8.4 per cent engineering and 16 per cent arts and social science in higher education with a trend in recent years for the numbers studying science and engineering to decline. Among middle managers in British industry 3 per cent had higher degrees, 18 per cent first degrees and 18 per cent diploma level qualifications. Of those with degrees 22 per cent studied arts subjects, 22 per cent social science, 20 per cent natural science and 36 per cent vocational subjects such as engineering and business studies. Typically, however, middle managers have attended local education authority grammar schools and then obtained further qualifications by part-time study. Among first-line managers, most left school at the age of 16, only some 10 per cent have more than an elementary or ordinary education, and only 16 per cent have any kind of qualification at all (Mansfield *et al.*, 1981; see also Leggatt, 1978; Melrose-Woodman, 1978; Poole and Mansfield, 1980; Torrington and Weightman, 1982).

The explanation for engineers' failure to penetrate the heart of the technical division of labour lies in the data just reviewed. The typical British manager is upwardly socially mobile and working in an industry or sector for which (even if he has received much formal education) he has not received any specific formal education and training. He works in an atmosphere of uncertainty and ambiguity. His job is often stressful and he receives little or no support from the information generated internally. The attractions of professional ideologies and management science can be enormous to many such people. They promise solutions to their problems and even if they cannot provide them, they may at least provide either *post hoc* justification for actions and a kind of respectability and psychological security (Glover, 1978b). The attractions of constructing rigid and rule-following bureaucratic modes of organization are also obvious, because they attempt to bring order and impose control on something which is ultimately unmanageable. While such managerial strategies are thus easily comprehensible they are not particularly beneficial to the engineer (cf. Burns and Stalker, 1961). While the engineer is also likely to be upwardly socially mobile he has an expertise which can provide integrity and a degree of certainty. He does not need either professionalism or bureaucracy to do his task. His skills however are the antithesis of the culture of many British organizations and consequently many engineers find it easier and more attractive to beat a path into management rather than to remain at the core of the technical division of labour where they belong, simply because that core is so derided and under-resourced in their country.

Conclusion

In this chapter we have once again emphasized the centrality of the engineers' tasks in developed societies, while noting how in Britain, at least, it appears to be insufficiently acknowledged for the mixture of social and political reasons discussed in Chapter 3 and elsewhere in this book. Both individually and collectively many engineers have variously responded to this unfortunate situation by attempting to identify themselves as scientists of a kind, as professionals or as managers. One important manifestation of this iatrogenic tendency, whereby the 'cure' for the disease only makes it worse, is the apparently still growing habit of describing engineering as technology and engineers as technologists. The irrelevance of the habit and its lack of integrity can be highlighted by pointing out that perhaps 90 per cent of working 'technologists' are engineers by qualification and job title. It can be observed in the usually status-concerned notion of 'high technology', with its apparent contempt for basic and normally crucial detail, and in some of its more ridiculous forms, in advertisements for the services of joiners and plumbers in which the relevant individuals describe themselves as 'woodologists' and 'plumologists'. Yet it cannot ever be stressed too strongly that the principal, defining, outputs of all types and levels of technical work are three-dimensional artefacts, not formal knowledge, as the use of the suffix '-ology' implies.

Of equal significance is the fact that any community in which there are massive disjunctions between the technical and social divisions of labour is normally one in which the most socially useful efforts are neither properly recognized nor properly rewarded. Such a community is likely to be turned in on itself, and severely hampered in its capacity to abolish the disjunction through an obsession with such immediate results as alienation, anomie, conflict, delusions and follies of grandeur, mis- and dis-organization, and various forms of material and intellectual poverty (including decadence). Yet no human situation is ever hopeless, and disaster and frustration are often the most positive seedcorn for positive change and regeneration. Some positive suggestions for change are contained in the next and final chapter.

12 *Conclusions*

Introduction

In this final chapter we first make some very brief points about the practical relevance of sociological knowledge to engineering. Then we extract, from the preceding chapters, some of what we think are the most important points in them. Taken together, these points are not a unified summary of everything in the book; rather they highlight things about which we feel more strongly. Finally we acknowledge some of the loose ends and gaps in our arguments and discuss some of their implications.

Although we do not believe that sociology offers superior explanations of economic, political and social phenomena than those of the other social sciences we think that its approaches tend to be more inclusive than theirs. Sociology's strengths and weaknesses reflect each other: breadth and comprehensiveness on one hand, flabby generalization on the other. Sociologists can be justly criticized for showing, in their work, prejudices common to many academics (or other intellectuals who only seem to know life from literature or from working in laboratories) who study life without much experience of it, and who sometimes put what they think are the interests of their subject above the search for truth. Yet the subject's breadth – implying comparisons between groups and societies simultaneously and over time – and its readiness to look under the surface of events, asking why things taken for granted are taken for granted, can be very powerful strengths indeed, and those who ignore its findings and insights can do so at their peril. People generalize anyway, and sociology can at least try to provide them with an effective aid and framework for doing so.

We hope that we have succeeded in showing how lay and sociological explanations of the problems of British engineering and British society can usefully inform each other. We believe that in doing so we have suggested some ways in which Britain's engineers can help themselves. Above all we hope that we have shown that engineers should *not* try to turn themselves into applied scientists, or professionals, and that they should instead stand up for and develop themselves on the basis of the reality of their own modes of working practice *as* engineers.

The Arguments So Far

In the first half of the book, where we defined our problem, we first argued that engineers and sociologists were both products of industrialization, who had developed separately but who might gain much from contact. In doing so we emphasized our idea of man as *homo faber*, the maker-and-doer cum engineer. We noted how classical sociology had been far from unsympathetic to this idea. We attributed the relatively low social position of the engineer in Britain to the whole evolution of British society since, and in some ways during, pre-industrial times. It was not some sort of historic anomaly but a major, even a distinguishing, characteristic of Britishness. It was reflected in the continued relative decline of Britain's economy *vis-à-vis* its major competitors over a period of 150 years. We argued that it was not, however, inevitable. Nor was it desirable for Britain or other countries. To opt out of the world's economic development was neither rational nor responsible.

In the second half of the book we entered into a fairly detailed sociological explanation of both particular and general aspects of the problem. We noted too the British systems of technical and other forms of education and training, while not originally at the root of engineering's difficulties, had certainly done much to make them worse since. Here educationalists had misclassified engineering as part of science, giving it a derivative, subordinate status. They had also helped to produce a situation in which engineering courses often attracted the wrong kind of talent, as well as an inadequate amount of it. We then looked at sociological and related discussions of motivation and job satisfaction, to show their variety and relevance to engineering. We argued that, in general, competent supervision meant putting the needs of tasks first and those of subordinates a close second; and in being aware both of the extent to which, and the ways in which, expectations of employees can vary.

In the next two chapters we offered relatively factual accounts of the work of some of the engineer's main colleagues and superiors, of the nature of major economic sectors and of public policies towards industry, as well as of the results of surveys of engineers' subjective experience of work. The latter, we argued, generally reflected the British tendency to exclude many engineers from management, which made them console themselves with interesting technical work and their home lives.

Our discussion of the collective organization of engineers argued that for engineers to belong to trade unions would often be a rational course of action, but that the lure of professionalism was something of an irrelevant distraction for most types of engineer. Engineering could not be a true profession because most of its members were salaried employees, not independent practitioners, and because their work sought to make and sell goods for the markets of the world, not to provide advice and other services to individual clients. Our final main chapter focused on what is in many ways the major concern of the book, the disjunction between the place of engineers in Britain in the technical and social divisions of labour. The technical role of engineers was central to productive and many related forms of activity. Engineering was central to the evolution or development of human societies. However the social position of Britain's engineers did not reflect this fact; they were regarded, employed, deployed and remunerated as if their abilities and tasks were significantly less important than they are. This generalization held even within many places of employment devoted to engineering, and in particular within and around the central activity of production in manufacturing.

Loose Ends and Implications

We have already made, or more often implied, a number of suggestions about the ways in which Britain's engineers might improve their position or condition. Our main original aim was to produce a sociological primer for students and teachers of engineering along with a 'managerial element', and some discussion of British engineers' special problems. We have done these in just about equal measure.

We have four main points and some lessers ones still to make. They do flow from what has gone before, if not very tidily. First, we both fear and hope that any unfavourable reviewers may put a title something like 'Yesterday's World' over their reviews of this book. Such a title would be a play on that of the BBC television programme 'Tomorrow's World' on the 'latest advanced in science and technology'; and the reviewer(s) might suggest that we seem to hanker back to the (to us) halcyon days of the mid-twentieth-century 'smoke-stack' or 'sunset' industries, and of large numbers working in grimy factories.

We are not ashamed to admit that we would like to see greater numbers employed, and few unemployed, as was the case between 1940 and the mid-1970s. Further we will come clean and admit that we do *like* manufacturing and most other places where engineers work, and that complicated artefacts fascinate us and that we believe that the typical engineer and the typical craftsman have jobs which must usually be at least as fulfilling as those of the administrator or routine clerical worker, and far more so in most instances. Much more seriously, however, we strongly distrust the whole idea of distinguishing between 'sunrise' and 'sunset' industries: an apocryphal German, Japanese or Swede keeps repeating to us that 'in *my* country, *all* the industries are sunrise industries!' (or at least 'daylight' ones). The idea that because something is old and in difficulty one abandons it for something more youthful and glamorous does not seem to us to be very responsible. Indeed it seems to combine fly-by-night and arms'-length attitudes very dangerously. Manufacturing sectors are in any case far too diverse, within as well as between themselves, in terms of outputs, processes, skills, satisfactions and performance for the facile sunrise–sunset distinction to mean much, except perhaps to lazy-minded journalists and politicians. It also seems a little arrogant to suggest, however politely, that factory work is somehow more suited to people in the developing countries than to Anglo-Saxons. It is also more than a little out of date and naive, given its increasing technical and intellectual sophistication. Also in our experience of factory and of other types of work (and we have a little of each), both routine clerical and menial service jobs, especially ones of the latter type which involve serving the public, are generally far more demeaning than either 'hard' or skilled manual ones.

Second, we have not written a great deal either about an 'information technology revolution' or 'science, technology and society'. The former term, we suspect, may by the year 2000 be about as respected as former prime minister Harold Wilson's 'the white heat of technology' of the early 1960s is today. Of course microprocessors offer a great deal for improving the efficiency of many kinds of task in saving labour, in saving money and time, in reliability, and so on. But the sensationalism of recent years about its impact on tasks and products (usually favourable) or on jobs (usually felt to be unfavourable) is not attractive. This is partly because the academics and journalists who have written about the so-called 'information revolution' are often hard-selling themselves, for what else have they ever had to offer but various kinds of information? It is partly because the British, whose industries seem particularly inept in adopting new processes, sometimes seem to have welcomed 'new technology' as a kind of cargo cult, whereby new crazes or fashions emanating from some mysterious external source hopefully distract attention from or possibly even solve pressing problems. We believe that it is very dangerous to suggest that problems of 'sunset' sectors can be forgotten or leap-frogged by embracing all that is rather snobbishly called 'hi-tech'. It is true that the British have been singularly good at natural scientific discovery, and sometimes at technical invention, compared with other countries. Yet apart from the fact there is little evidence to suggest that the two processes are often directly connected with each other in practice – rather than confused by academics – the British also remain singularly bad at manufacturing in most fields and no amount of doing the superficially harder or more novel things quite well is going to solve the very different and often much more basic original problems. Indeed they will normally be diversions from them.

On the notion that 'science, technology and society' might have been our focus, we think that we have written extensively and to at least some effect about all three, while sensibly repudiating the notion that they are related in the ways often implied when the (rather lordly) phrase is employed. We do not believe that engineering is best understood as scientists innocently discovering things which are then used by corrupt commercial interests in a black-box process called 'technology', or that the 'impact on society', the last word meaning the environment or employees of 'uncaring' multinationals, is normally bad. In other words we do not accept the definition of engineering as simply and only consisting of 'putting knowledge to use' (which in the European setting only the inhabitants of the British Isles seem to accept), and we do not see all or even a majority of its effects on employees, consumers or society in general as being harmful. Of course many are, and they should, indeed must, be the focus of serious study and rigorous action.

Some of the texts which adopt the above sort of approach towards 'science and society', and related phenomena, are often very imaginative, interesting and suggestive in technical, scientific, political and social terms (see Hales, 1982, for a particularly enjoyable example of the genre). Thus they look at technical and political issues in novel ways, and they undoubtedly do so with the interests of the majority of people at heart. Yet in spite of this we often find them confusing, in so far as they support the very technocratic (or rather intellectual) elitism which they are ostensibly written to attack, simply by using the pure–applied science view of technical phenomena. Critiques of this type seem to have a vested interest in their own intellectual complexity and contrariness.

Third, we are writing against a background of important and often

valuable changes in the content, as well as in the amount and organization, of technical and other forms of education and training. A danger
which we see here is one which we call 'neo-classicism'. The Oxford
Fallacy (Glover, 1979 and 1980), a veritable knee-jerk reaction of
products of traditional British (especially English and Welsh) education, asserts and assumes that a broad, civilizing education and a useful
and wealth-creating one cannot be mixed in the same person (see also
Wilkinson, 1964). In other words, it is argued that those classed as
'specialists' should not usually be given top jobs because they lack the
necessary breadth of vision. This argument is in fact an 'occupational
strategy' on the part of traditionally – and in fact quite narrowly-
educated – arts and natural science graduates, which aims to keep
them on top and people like engineers on tap. Not only is the argument illogical, it has been disproven for decades by the outputs of
continental European higher technical education (Ahlström, 1982;
Glover, 1978a). But its effects linger whenever technical parts or
streams in education and training are reserved for or left to the
'academically less able'. Even one or two prominent advocates of the
Royal Society of Arts' Education for Capability movement, which
aims to change society by supplementing education's emphasis on
critical understanding with an emphasis on practical and personal
competence, have been heard to express the view that Oxbridge
classics or history graduates, being very intelligent philosopher-kings
of a type, should still monopolize the *very* highest posts in public
service and government. Although we oppose this view very
strongly, we do however feel that no one, whatever his or her educational background, should be excluded from any kind or level of
employment if they have something useful to contribute. Further,
although we want to see engineering education expanded in both
concept and scale, we do not like the current tendency to think that it
should or can only be expanded at the expense of other ostensibly less
useful types of education.

Fourth, and following on from what we have just written as well as
from the rest of the book, we offer without comment our picture of
what a 'compleat engineer' for the twenty-first century might look
like. Because our depiction flows from what we have already said, any
further explanation or rationale would be superfluous:

> *Secondary Education:* the product should be very proficient
> numerically, verbally and orally; should have examination
> successes in two or three natural scientific subjects as well
> as mathematics, some social science, at least two or preferably
> three languages, and some strong sign of literary and artistic

interests as well as technical skills in areas such as typing, computing, metal working, design and woodwork. Thus secondary education should be broad, and with a fair amount of depth too in the more difficult subjects.

Undergraduate Education: breadth and depth as well as rigour in relevant scientific and economic subjects are basic; these should comprise the core upon which the learning of practical skills from very experienced practitioners ('engineer–professors') builds in the later parts of courses. These practical skills should be commercial, financial and socio-political as well as technical, and graduates should have been examined in such subjects as marketing, financial management and organizational behaviour. With regard to years of study the practical part of the course should be lengthy – preferably as long as the theoretical part and it should contain a genuine 'sandwich' element of bona fide employment. The degrees should be academically of masters standard and should normally take at least five years' study (six in Scotland). Courses should be as *sector-specific* (even employer-specific) as possible and their central underlying principle should be that the needs of tasks, *broadly* defined, should determine *everything*. This suggests the eventual demise of conventional and supposedly all-inclusive degrees especially in mechanical and to a smaller extent in electrical/electronic engineering. To do all of what is suggested it would probably help if students were to concentrate on specific technical packages (also involving commercial, financial and organizational issues) in the later years of their courses.

Personal Attributes: the products should generally be academically confident, able and outgoing students, capable of focused convergent thinking *as well as* of the more creative divergent kind. They should possess the 'helicopter quality' of being able to rise above problems, to be able to distinguish wood from trees quickly. High-powered and tough- and open-minded technically, commercially, financially, socially and politically, they should be potentially capable of usefully occupying senior posts in public service and government as well as in industry and commerce. This would be the norm, although there would be variations, to include less outgoing types and the 'salesperson' kind of engineer too.

Initial Employment Experience and Potential: the sandwich element of undergraduate studies should contain a mandatory element of shop-floor or equivalent experience, during which a 'proper job' should be competently performed. After graduation experience of production and of design and development work is desirable in the early years; with commercial and financial experience following very shortly thereafter, or even overlapping them.

A few other points remain to be made very briefly. First, we would suggest that engineers do need to involve themselves more often and more prominently in public affairs, whenever technical matters impinge on everyday life and wherever they feel that something is worth saying. This is a basic part of standing up for one's craft and experience. Engineers have allowed natural scientists to discuss engineering matters in public far too often; they sometimes have a right and indeed a duty to rebuke such acts of trespass. Engineers' representatives should certainly protest very loudly *indeed* when scientists are allowed to dominate public inquiries when engineering issues are to be dealt with.

Second, we do favour the somewhat hoary old idea of engineers having a protected title so that unqualified groups cannot describe themselves as engineers. This would and normally should not stop unqualified or partly qualified individuals from rising through the ranks and occupying posts normally occupied by the qualified. It would however stop skilled operatives seeming to put themselves on an educational par with graduates including holders of masters' degrees and doctorates. We are not arguing this point for snobbish reasons, but because all engineering is devalued when, for example, skilled operative members of prominent trade unions describe themselves as 'the engineers'.

Third, we feel, as many did at the time when it was being set up, that the Engineering Council should have been established as a statutory rather than as a chartered body. By being seen to be a creature of the state, and not just as one with state-delegated powers, it would probably have been in a much stronger position to determine the content of education and training. Also the British state would have been seen to have finally brought engineering into the centre of things.

Next, and most emphatically not as any kind of afterthought, we should comment upon the fact that 1984 was the Year of WISE – Women Into Science and Engineering. The best argument in favour of more women engineers was given us by the Scottish senior executive responsible for recruitment and training in a very large manufacturing company, that 'after all, they represent half the talent in the country', that is, engineering needs women. The worst argument is the opposite. Why (male? dominant? elitist?) '*Science and* (more intuitive? feminine? derivative? lower status?) *Engineering*?'

though? For those who would help engineering in Britain, the issues should be more important than slick acronyms and would-be trendy slogans, especially when they reflect the culture which they are meant to oppose.

Fifth and finally, the case against scientists, as opposed to engineers, at the top in manufacturing industry can be argued along the lines suggested in some of our earlier references to graduates in arts subjects. Knowledge of, and a liking for, the specifics of laboratory life – a couple of colleagues, long hours of watching things, the ability to control awkward variables – is not much of a preparation for 'managing industry'. A farmer might be likely to do a better job, if only because the risks he has to take are probably nearer to those of industry, than the 'risks' taken by the scientist.

One of our main themes, probably the major one, has been the fact that within manufacturing the core activity or function has been foolishly and eccentrically neglected in Britain; and within the country as a whole, manufacturing has been misassessed over the years, so that it has not provided a first-choice career for many potentially very competent people. Each effect has had an influence on, and has been informed by, the special status of engineering, and even the ways in which the British try to tackle 'the problem' turn out to be part of it. Much of the rhetoric of the day – of 'improving management quality', 'high technology', 'better industrial relations', 'corporate strategy', for example – is part of this. Similarly, sociological ideas uncritically put forward – (say) straight-Marx or straight-Weber – are also unlikely to do more than confuse the issues. Indeed they often have. But sociology can inform, and not only because a bit of Marx, a bit of Durkheim, a bit of Weber, and so on, belong to the rhetoric or usage which we have inherited for the purposes of discussing national ills. Marxian thought is misleading when it suggests that 'modern industry' and 'capitalist society' are not at all like anything that went before; we have witnessed much continuity as well as much change. Readers of second-hand accounts of Weber who conclude that 'rational behaviour' is a good thing because it is 'modern' are mistaken about Weber, about human nature, and about social change. Those who would misuse Durkheim's writings to support the notion that the division of labour is simultaneously 'good' and 'modern' are contradicted, at least in part, by the simpler and – in most senses of the phrase – less divisive approach to the design of jobs and organizations adopted by the Japanese and the West Germans.

What an appreciation of man as *homo faber* teaches us is that the springs and courses of human action are almost infinitely variable and difficult to predict. Sociology also suggests that there are however patterns which repeat themselves and trends which are followed. When sociology is used to study real-life events and when it is properly informed by awareness of both of these points it can, as we hope we have shown, shed some light on important issues. Certainly the difficulties facing engineering in Britain, indeed the whole issue of engineering in Britain, are important. We have not offered much in the way of broad solutions, partly because to do so was not part of our remit, partly because we would rather engineers and their other allies find their own (although for some educational and related suggestions see Glover, 1978a). What we do want to conclude by stressing, however, are two thoughts so far unwritten, but ones which we hope will have occurred at least once to all who have read so far. These are, simply, that if angels exist, they will generally be on the side of the engineer, and that heaven does often help those who help themselves.

References

Abernathy, W. J., Clark, K. B., and Kantrow, A. M. (1983), *Industrial Renaissance: Producing a Competitive Future for America* (New York: Basic Books).

Ahlström, G. (1982), *Engineers and Industrial Growth* (London: Croom Helm).

Albu, A. (1979), 'British attitudes to engineering education in a historical perspective', in K. Pavitt (ed.), *Technical Innovation and British Economic Performance* (London: Science Policy Research Unit, University of Sussex/ Macmillan).

Aldcroft, D. (1986), *The British Economy*, Vol. 1: *The Years of Turmoil 1920–51* (Brighton: Wheatsheaf).

Alderfer, C. P. (1972), *Existence, Relatedness and Growth: Human Needs in Organizational Settings* (New York: Free Press).

Allen, G. C. (1979), *The British Disease* (London: Institute of Economic Affairs).

Allen, R. W., Madison, D. L., Porter, L. W., Renwick, P. A., and Mayes, B. T. (1979), 'Organizational politics: tactics and characteristics of its actors', *California Management Review*, vol. 22, pp. 77–83.

Ardagh, J. (1973), *The New France* (Harmondsworth: Penguin).

Armstrong, M. (1984a), *A Handbook of Personnel Management Practice* (London: Kogan Page).

Armstong, P. (1984b), 'Competition between the organisational professions and the evolution of management control strategies', in K. Thompson (ed.), *Work Employment and Unemployment: Perspectives on Work and Society* (Milton Keynes: Open University Press).

Ashton, D. N. (1973), 'The transition from school to work', *Sociological Review*, vol. 21, 1 (February), pp. 101–25.

Ashton, D. N. (1974), 'Careers and commitment: the movement from school to work', in D. Field (ed.), *Social Psychology for Sociologists* (London: Nelson).

Ashton, D. N. and Field, D. (1976), *Young Workers: From School to Work* (London: Hutchinson).

Avineri, S. (1968), *The Social and Political Thought of Karl Marx* (Cambridge: Cambridge University Press).

Bacon, R., and Eltis, W. (1976), *Britain's Economic Problem: Too Few Producers* (London: Macmillan).

Badawy, M. K. (1978), 'One more time: how to motivate your engineers', *IEEE Transactions in Engineering Management*, vol. EM–25, no. 2 (May), pp. 37–42.

Baillie, K. (1974), *Institute of Personnel Management Membership Survey* (London: Institute of Personnel Management).

Bain, G. S. (1970), *The Growth of White Collar Unionism* (Oxford: Clarendon).

Bain, G. S. (ed.) (1983), *Industrial Relations in Britain* (Oxford: Blackwell).

Bain, G. S., Coates, D., and Ellis, V. (1973), *Social Stratification and Trade Unionism: A Critique* (London: Heinemann).

Baker, C. D. (1976), *Tort* (London: Sweet & Maxwell).

Balfour Committee (1929), *Board of Trade: Committee on Industry and Trade. Final Report* (Chairman: Arthur Balfour), Cmnd 3282 (London: HMSO).

Bamber, G. J. (1986), *Militant Managers? Managerial Unionism and Industrial Relations* (Aldershot: Gower).

Bamber, G. J., and Glover, I. A. (1975), *Study of the Steel Industry Management Association* (Edinburgh: Manpower Studies Research Unit, Heriot–Watt University).

Bank for International Settlements (1978), *Forty-Eighth Annual Report, 1 April 1977–31 March 1978* (Basle: Bank of International Settlements).

Barnett, C. (1972), *The Collapse of British Power* (London: Eyre Methuen).

Barnett, C. (1986), *The Audit of War* (London: Macmillan).

Bell, D. (1968), 'The measurement of knowledge and technology', in E. B. Shelton and W. E. Moore (eds), *Indicators of Social Change: Concepts and Measurements* (New York: Russell Sage Foundation).

Bell, D. (1978), *The Cultural Contradictions of Capitalism* (New York: Basic Books/Harper Colophon).

Bellini, J. (1981), *Rule Britannia: A Progress Report for Domesday 1986* (London: Cape).

Bendix, R. (1974), *Work and Authority in Industry: Ideologies of Management in the Course of Industrialization* (Berkeley, Calif.: University of California Press) (1st edn 1956).

Berg, I., with Gorelick, S. (1970), *Education and Jobs: The Great Training Robbery* (Harmondsworth: Penguin).

Berger, P. (ed.) (1964), *The Human Shape of Work: Studies in the Sociology of Occupations* (New York: Macmillan).

Berger, P., and Berger, B. (1976), *Sociology: A Biographical Approach* (Harmondsworth: Penguin) (1st edn 1972).

Berger, P., and Kellner, H. (1982), *Sociology Reinterpreted: An Essay on Method and Vocation* (Harmondsworth: Penguin).

Beuret, G., and Webb, A. (1983), *Goals of Engineering Education (GEEP): Engineers – Servants or Saviours?* (London: Council for National Academic Awards).

Blackaby, F. (ed.) (1978), *De-Industrialization* (London: Heinemann).

Blau, P. M., and Scott, W. R. (1963), *Formal Organizations: A Comparative Approach* (London: Routledge & Kegan Paul).

Blaug, M. (1965), 'The rate of return on investment in education in Great Britain', *Manchester School*, September 1965 (reprinted in Blaug, M. (ed.), 1968).

Blaug, M. (ed.) (1968), *The Economics of Education* (Harmondsworth: Penguin).

Blaug, M. (1972), *An Introduction to the Economics of Education* (Harmonds-worth: Penguin).

Blaug, M. (1974), *The Cambridge Revolution: Success or Failure?* (London: Institute of Economic Affairs).

Blauner, R. (1964), *Alienation and Freedom: The Factory Worker and His Industry* (Chicago and London: University of Chicago Press).

Bloch, M. (1962), *Feudal Society* , 2nd edn, (translated by L. A. Manyon) (London: Routledge & Kegan Paul).

Blondel, J. (1963), *Voters, Parties and Leaders* (Harmondsworth: Penguin).

Blumer, H. (1962), *Society as Symbolic Interaction*, in A. Rose (ed.), *Human Behaviour and Social Process* (London: Routledge & Kegan Paul).

Boltho, A. (1984), 'Economic policy and performance in Europe since the second oil shock', in M. Emerson (ed.), *Europe's Stagflation* (Oxford: Oxford University Press).

Braverman, H. (1974), *Labor and Monopoly Capital: The Degradation of Work in the Twentieth Century* (New York and London: Monthly Review Press).

Brewster, C. (1986), 'A typology of managerial controls', paper presented to the Fourth Aston/UMIST International Conference on the Labour Process (Birmingham: University of Aston).

Briggs, A. (1959), *The Age of Improvement* (London: Longman).

British Institute of Management (BIM) (1976), *Front-Line Management* (London: BIM).

Brittain, S. (1981), 'America falls victim to the British disease', *Financial Times* (20 March).

Broom, L., and Selznick, P. (1977), *Sociology: A Text with Adapted Readings* (6th edn) (New York: Harper & Row).

Brown, D., and Harrison, M. J. (1978), *A Sociology of Industrialization: An Introduction* (London: Macmillan).

Brown, R. (1976), 'Women as employees: some comments on research in industrial sociology', in D. L. Barker and S. Allen (eds), *Dependence and Exploitation in Work and Marriage* (London: Longman).

Brown, W. (ed.) (1981), *The Changing Contours of British Industrial Relations: A Survey of Manufacturing Industry* (Oxford: Blackwell).

Buchanan, R. A. (1985), 'Institutional proliferation in the British engineering profession, 1847–1914', *Economic History Review*, vol. 39, pp. 42—60.

Burgess, T., and Pratt, J. (1970), *Policy and Practice: The Colleges of Advanced Technology* (Harmondsworth: Allen Lane).

Burn, W. L. (1964), *The Age of Equipoise: A Study of the Mid-Victorian Generation* (London: Allen & Unwin).

Burns, T., and Stalker, G. M. (1961), *The Management of Innovation* (London: Tavistock).

Business Graduates Association (1977), *Higher Management Education and the Production Function* (London: BGA).

Cairncross, A. (1978), 'What is de-industrialization?', in F. Blackaby (ed.), *De-Industrialization* (London: Heinemann).

Caves, R. E., and Krause, L. B. (1980), *Britain's Economic Performance* (Washington DC: Brookings Institution).

Chapman, R. A. (1968), 'Profile of a profession: the administrative class of the Civil Service', in *The Civil Service*, Vol. 3 (2) *Surveys and Investigations: Evidence Submitted to the Committee Under the Chairmanship of Lord Fulton 1966–8* (London: HMSO).

Chapman, R. A. (1970), *The Higher Civil Service in Britain* (London: Constable).

Charlesworth, J., and Morse, G. (1983), *Principles of Company Law* (12th edn (London: Stevens).

Charlesworth, J., Schmitthof, C. M., and Sarre, D. A. G. (1984), *Mercantile Law* (14th edn) (London: Stevens).

Child, J. (1978), 'The non-productive component within the productive sector', in M. Fores and I. Glover (eds), *Manufacturing and Management* (London: HMSO).

Child, J. (1982), 'Professionals in the corporate world: values, interests and control', in D. Dunkerley and G. Salaman (eds), *International Yearbook of Organization Studies* (London: Routledge & Kegan Paul).

Child, J. (1984), *Organization: A Guide to Problems and Practice* (2nd edn) (New York: Harper & Row).

Child, J., and Ellis, T. (1973), 'Predictors of variation in managerial roles', *Human Relations*, vol. 26, no. 2 (April), pp. 227–50.

Child, J., Fores, M., Glover, I., and Lawrence, P. (1983), ' A price to pay? Professionalism and work organisation in Britain and West Germany, *Sociology*, vol. 17, no. 1, pp. 63–78.

Child, J., Fores, M., Glover, I., and Lawrence, P. (1986), 'Professionalism and work organization: reply to McCormick', *Sociology*, vol. 20 no. 4, pp. 49–55.

Child, J., and Partridge, B. (1982), *Lost Managers: Supervisors in Industry and Society* (Cambridge: Cambridge University Press).

Clegg, H. A. (1979), *The Changing System of Industrial Relations in Great Britain* (Oxford: Blackwell).

Collingridge, D. (1981), *Social Control of Technology* (Oxford: Oxford University Press).

Conservative Political Centre (1978), *The Engineering Profession: A National Investment* (London: Conservative Political Centre).

Cotgrove, S. F. (1958), *Technical Education and Social Change* (London: Allen & Unwin).

Cotgrove, S. (1978), *The Science of Society: An Introduction to Sociology* (London: Allen & Unwin).

Cotgrove, S. (1982), *Catastrophe or Cornucopia: the Environment, Politics and the Future* (New York: Wiley).

Cotgrove, S., and Box, S. (1970), *Science Industry and Society: Studies in the Sociology of Science* (London: Allen & Unwin).

Cunningham, P., and Turnbull, M. (1981), *International Marketing and Purchasing* (London: Macmillan, 1981).

Dahl, R. A. (1963), *Modern Political Analysis* (Englewood Cliffs, NJ: Prentice-Hall).

Dahrendorf, R. (1959), *Class and Class Conflict in an Industrial Society* (London: Routledge & Kegan Paul) (first published in Germany 1957).

Dainton Report (1968), *Report of an Inquiry into the Flow of Candidates in Science and Technology into Higher Education*, Cmnd 3541 (London: HMSO).

Dale, H. E. (1941), *The Higher Civil Service of Great Britain* (Oxford: Oxford University Press).

Devonshire Commission (1871–5), *Report of the Royal Commission on Scientific Instruction and the Advancement of Science*, Parliamentary Papers 1871, vol. 24; 1872, vol. 25; 1873, vol. 28; 1874, vol. 22; 1875, vol. 28.

Dickens, L. (1972), 'UKAPE: a study of a professional union', *Industrial Relations Journal*, vol. 3, pp. 3–16.

Diggins, J. P. (1978), *The Bard of Savagery: Thorstein Veblen and Modern Social Theory* (New York: Seabury Press).

Dixon, T. J. (1980), 'The Civil Service syndrome', *Management Today* (May), pp. 74–9, 154, 162.

Dobson, A. (1984), *Sale of Goods and Consumer Credit* (London: Sweet & Maxwell).

Dodd, C. H. (1967), 'Recruitment to the administrative class 1960–1964', *Public Administration*, vol. 45, pp. 55–80.

Donovan Commission (1968), *Report of the Royal Commission on Trade Unions and Employers' Associations*, Cmnd 3623 (Chairman Lord Donovan) (London: HMSO).

Dore, R. (1973), *British Factory – Japanese Factory* (London: Allen & Unwin).

Dore, R. (1976), *The Diploma Disease: Education, Qualification and Development* (London: Allen & Unwin).

Drucker, P. F. (1970), *Technology, Management and Society* (London: Heinemann).

Drucker, P. F. (1981), 'The coming rediscovery of scientific management', in P. F. Drucker (ed.), *Toward the Next Economics and Other Essays* (London: Heinemann).

Dunnett, P. J. S. (1980), *The Decline of the British Motor Industry: The Effects of Government Policy 1945–79* (London: Croom Helm).

Durkheim, E. (1933), *The Division of Labour in Society* (translated by G. Simpson) (New York: Free Press).

Durkheim, E. (1952), *Suicide: A Study in Sociology* (translated by J. A. Spaulding and G. Simpson) (London: Routledge & Kegan Paul).

Edwardes, M. (1978), 'British industry and industry in Britain', in M. Fores and I. Glover (eds), *Manufacturing and Management* (London: HMSO).

Edwards, R. C. (1979), *Contested Terrain: The Transformation of the Workplace in the Twentieth Century* (New York: Basic Books).

Eldridge, J. E. T. (1971), *Sociology and Industrial Life* (London: Michael Joseph),

Elias, N. (1978), *What is Sociology?* (translated by Stephen Mennell and Grace Morrissey: foreword by Reinhard Bendix) (London: Hutchinson).

Elkin, F. and Handel, G. (1972), *The Child and Society: The Process of Socialization* (New York: Random House).

Ellul, J. (1964), *The Technological Society* (London: Cape).

Employment Act 1980, *Elizabeth II, 1980*, ch. 42 (London: HMSO).

Employment Act 1982, *Elizabeth II, 1982*, ch. 46 (London: HMSO).

Employment Protection Act 1975, *Elizabeth II, 1975*, ch. 71 (London: HMSO).

Engineering Council (1983), *The 1983 Survey of Professional Engineers*, (London: Engineering Council).

Engineers' Registration Board (1982), *Engineers' Registration Board* (London: Council of Engineering Institutions).

Ensor, R. C. K. (1946), *England 1870–1914* (Oxford: Clarendon).

Equal Pay Act 1970, *Elizabeth II, 1970*, ch. 41 (London: HMSO).

Esland, G. and Salaman, G. (eds) (1978), *The Politics of Work and Occupations* (Milton Keynes: Open University).

Fair Trading Act 1973, *Elizabeth II, 1973*, ch. 41 (London: HMSO).

Faulkner, A. C. and Wearne, S. H. (1979), *Professional Engineers' Needs for Managerial Skills and Expertise: Report of a Survey* (Bradford: University of Bradford School of Technological Management).

Feldman, A. S. and Moore, W. E. (1965), 'Are industrial societies becoming alike?', in A. W. Gouldner and S. M. Miller (eds), *Applied Sociology* (New York: Free Press).

Fidler, J. (1981), *The British Business Elite* (London: Routledge & Kegan Paul).

Field, D. (ed) (1974), *Social Psychology for Sociologists* (London: Nelson).

Field, D. (1982), *Inside Employment Law* (London: Pan).

Fielden, G. B. R. (Chairman) (1963), *Engineering Design* (London: HMSO).

Finniston, H. M. (Chairman) (1980), *Engineering Our Future: Report of the Committee of Inquiry into the Engineering Profession*, Cmnd 7794 (London: HMSO).

Finniston, Sir H. M. (1984), 'Overview of issues in engineering education', in S. Goodlad (ed.), *Education for the Professions* (London: SRHE and NFER-Nelson).

Fores, M. (1972), 'Engineering and the British Economic Problem', *Quest*, no. 22 (Autumn 1972), pp. 21–3.

Fores, M. (1977), *Scientists on Technology: A Confusing Saga*, Occasional Paper No. 571 (London: Free University of Kensington).

Fores, M. (1979), 'The myth of technology and industrial science', Discussion Paper 79–49 (Berlin: International Institute of Management).

Fores, M. (1981), 'The myth of a British industrial revolution', *History* (June), pp. 181–98.

Fores, M. (1982a), 'Homo faber and the American disease', *Cambridge Review*, pp. 241–7.

Fores, M. (1982b), ' "New information technology" as another "industrial revolution": teacher gets it wrong', Discussion Paper 82–6 (Berlin: International Institute of Management).

Fores, M. J. and Bongers, N. (1976), *The Engineer in Western Europe* (London: Department of Industry).

Fores, M., and Clark, D. (1975), 'Why Sweden manages better', *Management Today*, February, pp. 66–9.

Fores, M., and Glover, I. A. (1975), *Project 2735: The Mid-Victorian Construction of the British Problem: Introduction*, Occasional Paper No. 271 (London: Free University of Kensington).

Fores, M., and Glover, I. A. (1976a), 'Engineers in France', *Chartered Mechanical Engineer*, April, pp. 77–80.

Fores, M., and Glover, I. (1976b) 'The real work of executives', *Management Today*, November, pp. 104–8.

Fores, M., and Glover, I. (eds) (1978a), *Manufacturing and Management* (London: HMSO).

Fores, M., and Glover, I. (1978b), 'The British disease: professionalism', *The Times Higher Education Supplement*, 24 February.

Fores, M. and Pratt, J. (1980), 'Engineering: our last chance', *Higher Education Review*, vol. 12, no. 3, pp. 5–26.

Fores, M., and Rey, L. (1979), *'Technik:* the relevance of a missing concept', *Higher Education Review*, vol. 11, Spring.

Fores, M., and Sorge, A. (1981), 'The decline of the management ethic', *Journal of General Management*, vol. 6, no. 3, pp. 36–50.

Foy, N. (1978), *The Missing Links: British Management Education in the Eighties*, prepared for the Foundation for Management Education (Oxford: Oxford Centre for Management Studies).

Francis, A. (1980), 'Families, firms and finance capital', *Sociology*, vol. 14, no. 1, pp. 1–27.

Frank, A. G. (1980), *Crisis: In The World Economy* (London: Heinemann).

French, J. R. P. Jr, and Raven, B. (1968), 'The bases of social power', in D. Cartwright and A. Zander (eds), *Group Dynamics* (New York: Harper & Row).

Galbraith, J. K. (1967), *The New Industrial State* (London: Hamish Hamilton).

Gallagher, C. C. (1980), 'The history of batch production and functional factory layout', *The Chartered Mechanical Engineer*, vol. 27, no. 4, pp. 73–6.

Gannicot, K., and Blaug, M. (1969), 'Manpower forecasting since Robbins: a science lobby in action', *Higher Education Review*, vol. 2, no. 1, pp. 56–74.

Gerstl, J. E., and Hutton, S. P. (1966), *Engineers: The Anatomy of a Profession* (London: Tavistock).

Giddens, A. (1973), *The Class Structure of the Advanced Societies* (London: Hutchinson).

Giddens, A. (1981), *The Class Structure of the Advanced Studies* (revised edn) (New York: Harper & Row).

Giddens, A. and Held, D. (1982), *Classes, Power and Conflict: Classical and Contemporary Debates* (London: Macmillan).

Giddens, A., and Mackenzie, G. (1982), *Social Class and the Division of Labour: Essays in Honour of Ilya Neustadt* (Cambridge: Cambridge University Press).

Gill, R., and Lockyer, K. (1979), *The Career Development of the Production Manager in British Industry* (London: British Institute of Management).

Gill, C., Morris, R. S., and Eaton, J. (1977), 'APST: the rise of a professional union', *Industrial Relations Journal*, vol. 8, no. 1., pp. 50–61.

Gill, C., Morris, R., and Eaton, J. (1978), *Industrial Relations in the Chemical Industry* (Farnborough: Saxon House).

Glaser, B. G., and Strauss, A. L. (1968), *The Discovery of Grounded Theory: Strategies for Qualitative Research* (London: Wiedenfeld & Nicolson).

Glover, I. (1973), *The Sociological and Industrial Relations Literature on British Professional Engineers and Engineering* (London: City University).

Glover, I. A. (1974), *The Backgrounds of British Managers: A Review of the Evidence* (Edinburgh: Department of Industry/Manpower Studies Research Unit, Heriot-Watt University).

Glover, I. (1975), 'Engineering out in the British industrial cold', *Guardian*, 17 October, p. 17.

Glover, I. A. (1977a), *Managerial Work: A Review of the Evidence* (London: Department of Industry/City University).

Glover, I. A. (1977b), *The Economic Standing of the Engineering Profession* (London: Report to the Engineers' and Managers' Association).

Glover, I. A. (1978a), 'Executive career patterns: Britain, France, Germany and Sweden', in M. Fores and I. Glover (eds), *Manufacturing and Management* (London: HMSO).

Glover, I. (1978b), 'Professionalism and manufacturing industry', in M. Fores and I. Glover (eds), *Manufacturing and Management* (London: HMSO).

Glover, I. A. (1979), *Managerial Work: The Social Scientific Evidence and its Character*, unpublished PhD thesis (London: City University).

Glover, I. (1980), 'Social science, engineering and society', *Higher Education Review*, vol. 12, no. 3, pp. 27–42.

Glover, I. A. (1985), 'How the West was lost? Decline of engineering and manufacturing in Britain and the United States', *Higher Education Review*, vol. 17, no. 3, pp. 3–34.

Glover, I. A., and Fores, M. J. (1973), 'Engineers in Sweden', *Chartered Mechanical Engineer*, vol. 20, no. 11, pp. 80–3.

Glover, I. A., and Garbutt, D. (1986), 'The backgrounds of British managers: two hitherto unpublished studies and an update', mimeo, Dundee College of Technology.

Glover, I., and Herriot, P. (1982), 'Engineering students and manufacturing industry: chalk and cheese?', *Energy World*, April, pp. 8–12.

Glover, I. A., and Martin, G. (1986), 'Managerial work: an empirical and cultural contradiction in terms', paper presented at the Annual Conference of the British Sociological Association, University of Loughborough.

Glover, I., and Schröck, R. (1983), 'Same ingredients, different package?', *The Times Higher Education Supplement*, 23 September, p. 12.

Goffman, E. (1952), 'On cooling the mark out', *Psychiatry*, vol. 15, pp. 451–63.

Goldthorpe, J. H. (1959), 'Technical organization as a factor in supervisor–worker conflict: some preliminary observations on a study made in the mining industry', *British Journal of Sociology*, vol. 10, pp. 213–30.

Goldthorpe, J. H. (1966), 'Attitudes and behaviour of car assembly workers: a deviant case and a theoretical critique', *British Journal of Sociology*, vol. 17, pp. 227–44.

Goldthorpe, J. H., and Lockwood, D. (1963), 'Affluence and the British class structure', *Sociological Review*, vol. 11, pp. 133–63.

Goldthorpe, J. H., Lockwood, D., Bechhofer, F., Platt, J. (1969), *The Affluent Worker: Political Attitudes and Behaviour* (Cambridge: Cambridge University Press).

Goldthorpe, J., Lockwood, D., Bechhofer, F., Platt, J. (1969), *The Affluent Worker in the Class Structure* (Cambridge: Cambridge University Press).

Goldthorpe, J., Lockwood, D., Bechhofer, F., and Platt, J. (1970), *The Affluent Worker: Industrial Attitudes and Behaviour* (Cambridge: Cambridge University Press).

Gomulka, S. (1979), 'Britain's slow industrial growth: increasing inefficiency versus low rate of technological change', in W. Beckerman (ed), *Slow Growth in Britain* (Oxford: Oxford University Press).

Gospel, H. F. (1973), 'An approach to a theory of the firm in industrial relations', *British Journal of Industrial Relations*, pp. 211–28.

Gouldner, A. W. (1957–8), 'Cosmopolitans and locals: towards an analysis of latent social roles', *Administrative Science Quarterly*, Dec–March, pp. 281–306, 444–80.

Gowing, M. (1978), 'Science, technology and education: England in 1870', *Oxford Review of Education*, 4, pp. 3–17.

Granick, D. (1962), *The European Executive* (New York: Doubleday).

Guha, A. S. (1981), *An Evolutionary View of Economic Growth* (Oxford: Clarendon).

HMSO (1984), *Britain 1984: An Official Handbook* (London: HMSO).

HMSO (1986), *Britain 1986: An Official Handbook* (London: HMSO).

Habermas, J. (1971), *Towards a Rational Society*, translated by J. J. Shapiro (London: Heinemann) (first published in German 1968).

Hales, M. (1982), *Science or Society? The Politics of the Work of Scientists* (London: Pan).

Halsey, A. H., and Trow, M. (1971), *The British Academics* (London: Faber).

Hampden-Turner, C. (1983), *Gentleman and Tradesmen: The Values of Economic Catastrophe* (London: Routledge & Kegan Paul).

Haralambos, M., with Heald, R. M. (1980), *Sociology: Themes and Perspectives* (Slough: University Tutorial Press, 1980).

Harbury, C., and Lipsey, R. G. (1983), *An Introduction to the UK Economy* (London: Pitman).

Harvey, B. W. (1982), *The Law of Consumer Protection and Fair Trading* (2nd edn) (London: Butterworth).

Hawkins, K. (1978), *The Management of Industrial Relations* (Harmondsworth: Penguin).

Hayek, F. (1955), *The Counter-Revolution of Science* (Glencoe: Free Press).

Hayes, R. H. (1981), 'Why Japanese factories work', *Harvard Business Review*, July–Aug., pp. 57–66.

Health and Safety at Work . . . Act 1974, *Elizabeth II, 1974*, ch. 37 (London: HMSO).

Hepple, R. (1983), 'Industrial labour law', in G. S. Bain (ed), *Industrial Relations in Britain* (Oxford: Blackwell).

Herriot, P., Ecob, R., and Glover, I. (1981), 'Students' intentions and job types – fit for an intellectual? Engineering undergraduates' attitudes towards production and other functions of manufacturing units', *European Journal of Engineering Education*, vol. 5, pp. 301–8.

Herriot, P., Ecob, J. R., Hutchinson, M. (1980), 'Pure scientists or sordid salesmen?', *The Chartered Mechanical Engineer*, July, pp. 69–70.

Herriot, P., and Rothwell, C. (1981), 'Organizational choice and decision theory: effects of employers' literature and selection interview', *Journal of Occupational Psychology*, vol. 54, pp. 17–31.

Herzberg, F. (1966), *Work and the Nature of Man* (New York: World Publishing Co.).

Herzberg, F., Mausner, B., and Snyderman, B. (1959), *The Motivation to Work* (2nd edn) (New York: Wiley).

Hill, S. (1981), *Competition and Control at Work: The New Industrial Sociology* (London: Heinemann).

Hinton, Lord (1970), *Engineers and Engineering* (Oxford: Oxford University Press).

Hislop, M. (1971), 'The industry and the market', in E. F. L. Brech (ed.), *Construction Management in Principle and Practice* (London: Longman).

Holloway, S. W. F. (1964), 'Medical education in England 1830–1858: a sociological analysis', *History*, vol. 49, pp. 299–323.

Horne, D. (1969), *God is An Englishman* (Harmondsworth: Penguin).

Horton, J. (1964), 'The dehumanization of alienation and anomie: a problem in the ideology of sociology', *British Journal of Sociology*, vol. 15, pp. 283–300.

Hudson, L. (1966), *Contrary Imaginations: a Psychological Study of the English Schoolboy* (London: Methuen).

Hudson, L. (1968), *Frames of Mind: Ability, Perception and Self-Perception in the Arts and Sciences* (London: Methuen).

Hudson, L. (1975), *Human Beings* (London: Cape).

Hudson, L. (1978), 'Making things', in M. Fores and I. Glover (eds), *Manufacturing and Management* (London: HMSO).

Hughes, H. S. (1968), *Consciousness and Society: The Reorientation of European Social Thought 1890–1930* (New York: Knopf).

Hunter, S. L. (1972), *The Scottish Education System* (Oxford: Pergamon).

Hurd, G. (ed.) (1986), *Human Societies: A Sociological Introduction* (London: Routledge & Kegan Paul).

Hutton, S. P., and Lawrence, P. A. (1981), *German Engineers: the Anatomy of a Profession* (Oxford: Oxford University Press).

Hyman, R., and Price, R. (1983), *The New Working Class? White Collar Workers and Their Organizations* (London: Macmillan).

Industrial Relations Act 1971, *Elizabeth II, 1971*, ch. 72 (London: HMSO).

Ingham, G. (1984), *Capitalism Divided? City and Industry in British Social Development* (London: Macmillan).

Jackson, M. P. (1982), *Industrial Relations: a Textbook* (London: Croom Helm) (1st edn, 1977).

Jackson, M. P. (1982), *Trade Unions* (London: Longman).

Jamieson, I., and Lightfoot, M. (1982), *Schools and Industry: Derivations from the Schools Council Industry Project,* Schools Councils Working Paper No. 73 (London: Methuen).

Jenkins, D. (1978), 'The supervisor solution', *Management Today*, May, pp. 75–147.

Jewkes, J., Sawers, D., and Stillerman, R. (1969), *The Sources of Invention* (London: Macmillan) (1st edn published 1958).

Johnson, T. J. (1972), *Professions and Power* (London: Macmillan).

Jones, F. E. (Chairman) (1967), *Brain Drain: Report of the Working Group on Manpower for Scientific Growth,* Cmnd 3417 (London: HMSO).

Kaldor, Lord (1984), 'Wasted on an overseas posting: North Sea Oil should have been used to boost Britain's manufacturing capacity', *Guardian*, Agenda article based on speech to the House of Lords, 2 May.

Kelly, M. P. (1980), *White-Collar Proletariat: The Industrial Behaviour of British Civil Servants* (London: Routledge & Kegan Paul).

Kelly, M. P. (1983), 'A study of white-collar unions in dispute', *Industrial Relations Journal*, vol. 14, pp. 43–55.

Kelly, M. P., Martin, G., and Pemble, R. J. (1984), 'Unofficial action in white collar unions: the East of Scotland dimension in the 1981 Civil Service pay campaign', *Employee Relations*, vol. 6, pp. 9–16.

Kelsall, R. K. (1955), *Higher Civil Servants in Britain from 1870 to the Present Day* (London: Routledge & Kegan Paul).

Kelsall, R. K. (1974), 'Recruitment to the Higher Civil Service: how has the pattern changed?', in P. Stanworth and A. Giddens (eds), *Elites and Power in British Society* (Cambridge: Cambridge University Press).

Kerr, C., Dunlop, J. T., Harbison, F. H., and Myers, C. A. (1973), *Industrialism and Industrial Man: The Problems of Labour and Management in Economic Growth*, with a foreword by Haddon, R. (Harmondsworth: Penguin). (2nd edn; 1st edn Harvard University Press, 1960).

Kirby, M. (1981), *The Decline of British Economic Power since 1870* (London: Allen & Unwin).

Klemm, F. (1959), *A History of Western Technology* (London: Allen & Unwin).

Kogon, E. (1976), *Die Stunde der Ingenieure* (Dusseldorf, VDI Verlag).

Kotter, J. P. (1977), 'Power, dependence and effective management', *Harvard Business Review*, July–Aug., pp. 125–36.

Kotter, J. P. (1982a), 'What effective managers really do', *Harvard Business Review*, Nov.–Dec., pp. 156–67.

Kotter, J. P. (1982b), *The General Managers* (New York: Free Press).

Kuhn, T. S. (1970), *The Structure of Scientific Revolutions*, (2nd edn) (Chicago: University of Chicago Press) (1st edn 1962).

Kumar, K. (1978), *Prophecy and Progress: The Sociology of Industrial and Post-Industrial Society* (Harmondsworth: Penguin).

Landes, D. S. (1969), *The Unbound Prometheus: Technological Change and Industrial Development in Western Europe from 1750 to the Present* (Cambridge: Cambridge University Press).

Langford, D. A. (1982) *Direct Labour Organisations in the Construction Industry* (Aldershot: Gower).

Langrish, J., Gibbons, M., Evans, W. G., and Jevons, F. R. (1972), *Wealth from Knowledge: A Study of Innovation in Industry* (London: Macmillan).

Lawrence, P. A. (1977), 'The engineer and society, *Energy World*, June, pp. 2–4.

Lawrence, P. A. (1980), *Managers and Management in West Germany* (London: Croom Helm).

Lawrence, P. A. (1982), *Swedish Management: Context and Character*, report to the Social Science Research Council, December.

Lawrence, P. A. (1983), 'The United Kingdom as industrial periphery', paper presented to the British Sociological Association Annual Conference.

Lawrence, P. A. (1986), *Invitation to Management* (Oxford: Blackwell).

Lawrence, P. A., and Lee, R. A. (1984), *Insight into Management* (Oxford: Oxford University Press).

Leach, G. (1965), 'Technophobia on the left: are British intellectuals anti-science?', *New Statesman*, vol. 70, 27 August.

Leggatt, T. (1978), 'Managers in industry: their background and education', *Sociological Review*, vol. 26, pp. 807–25.

Lenski, G., and Lenski, J. (1978), *Human Societies: An Introduction to Macrosociology* (3rd edn) (New York: McGraw Hill).

Levy, J. C. (1983), 'Standards and routes to registration (1985 onwards). A paper for discussion' (London: Engineering Council–Professional Institutions Directorate).

Lewis, D. (1983), *Essentials of Employment Law* (London: Institute of Personnel Management).

Limprecht, J. A., and Hayes, R. H. (1982), 'Germany's world class manufacturers', *Harvard Business Review*, Nov.–Dec., pp. 137–45.

Littler, C. R. (1982), *The Development of the Labour Process in Capitalist Societies* (London: Heinemann).

0

Littler, C. R., and Salaman, G. (1984), *Class at Work: The Design, Allocation and Control of Jobs* (London: Batsford).

Lockwood, D. (1958), *The Blackcoated Worker* (London: Allen & Unwin).

Lockwood, D. (1966), 'Sources of variation in working-class images of society', *Sociological Review*, vol. 14, pp. 249–67.

Lorenz, K. Z. (1937), 'Imprinting', *The AUK*, vol. 54, pp. 245–73.

Loveridge, R. (1983), 'Sources of diversity in internal labour markets', *Sociology*, vol. 17, pp. 44–62.

McCormick, K. (1979), 'Manpower forecasters as lobbyists: a case study of the working group on manpower parameters for scientific growth 1965–8', in T. Whiston (ed.), *The Uses and Abuses of Forecasting* (London: Macmillan).

Macfarlane, A. (1978), *The Origins of English Individualism: the Family, Property and Social Transition* (Oxford: Blackwell).

McGregor, D. (1960), *The Human Side of Enterprise* (New York: McGraw Hill).

McQuillan, M. K. (1978), *Graduate Engineers in Production* (Cranfield: Cranfield Institute of Technology).

Maddison, A. (1979), 'The long run dynamics of productivity growth', in W. Beckerman (ed.), *Slow Growth in Britain* (Oxford: Oxford University Press).

Maitland, I. (1980), 'Disorder in the British workplace: the limits of consensus', *British Journal of Industrial Relations*, vol. 18, pp. 353–64.

Maitland, I. (1983), *The Causes of Industrial Disorder: a Comparison of a British and a German Factory* (London: Routledge & Kegan Paul).

Mansfield, R., Poole, M., Blyton, P., Frost, P. (1981), 'The British manager in profile', *Management Survey Report No. 5* (London: British Institute of Management).

Mant, A. D. (1977), *The Rise and Fall of the British Manager* (London: Macmillan).

Marcuse, H. (1964), *One Dimensional Man* (London: Routledge & Kegan Paul).

Marengo, F. D. (1979), *The Code of British Trade Union Behaviour* (Farnborough: Saxon House).

Marshall, E. (1982), *General Principles of Scots Law* (London: Green).

Marshall, E. A. (1983), *Scots Mercantile Law* (London: Green).

Martin, G., Frenkel, S., and Glover, I. (1987), 'Constructing organisations: a study of organisational design and change in the construction industry', in J. McGoldrick (ed.), *Business Case File in Behavioural Science* (Wokingham: Van Nostrand Reinhold).

Maslow, A. H. (1943), 'A theory of human motivation', *Psychological Review*, vol. 50, pp. 370–96.

Maslow, A. H. (1954), *Motivation and Personality* (New York: Harper).

Mathias, P. (1983), *The First Industrial Nation: An Economic History of Britain 1700–1914* (2nd edn) (London: Methuen).

Mead, G. H. (1934), *Mind, Self and Society* (Chicago: University of Chicago Press).

Mechanic, D. (1962), 'Sources of power of lower-level participants in complex organisations', *Administrative Science Quarterly*, vol. 7, pp. 349–64.

Melrose-Woodman, J. (1978), 'Profile of the British manager', *Management Survey Report No. 38* (London: British Institute of Management).

Merton, R. K. (1957), *Social Theory and Social Structure* (revised edn) (New York: Free Press; London: Collier Macmillan (1st edn 1949).

Merton, R. K., Gray, A. P., Hockey, B., and Selvin, H. (eds) (1952), *Reader in Bureaucracy* (New York: Free Press; London: Collier Macmillan).

Miliband, R. (1969), *The State in Capitalist Society* (London: Weidenfeld & Nicolson).

Miners, H. (1982), 'How staff jobs weaken line management', *Management Today*, June, pp. 33–40.

Mintzberg, H. (1973), *The Nature of Managerial Work* (Englewood Cliffs, NJ: Prentice-Hall).

Mintzberg, H., Raisinghani, D., and Théorêt, A. (1976), 'The structure of "unstructured" decision processes', *Administrative Science Quarterly*, vol. 21, pp. 246–75.

Monck, B. (1954), *The Status of the Engineer in Management* (London: British Association for the Advancement of Science).

Moore, B. (1967), *Social Origins of Dictatorship and Democracy: Lord and Peasant in the Making of the Modern World* (Harmondsworth: Allen Lane).

Mouzelis, N. (1967), *Organization and Bureaucracy: An Analysis of Modern Theories* (London: Routledge & Kegan Paul).

Mumford, L. (1966), 'Technics and the nature of man', *Technology and Culture* vol. 7, pp. 303–17.

National Economic Development Council/Manpower Services Commission (1984), *Competence and Competition: Training and Education in the Federal Republic of Germany, the United States and Japan* (London: National Economic Development Council/Manpower Services Commission).

National Institute Economic Review (various years) (London: National Institute of Economic and Social Research).

Neustadt, I. (1965), *Teaching Sociology: an Inaugural Lecture* (Leicester: Leicester University Press).

New, C. (1976), *Managing Manufacturing Operations*, BIM Management Survey Report No. 35 (London: British Institute of Management).

Nicholls, T. (1969), *Ownership, Control and Ideology* (London: Allen & Unwin).

Nyman, S., and Silberston, A. (1978), 'The ownership and control of industry', *Oxford Economic Papers*, vol. 30, 1, pp. 74–101.

OECD (1970), *Occupational Structure of the Labour Force and Economic Development* (Paris: OECD).

Olson, M., Jnr (1965), *The Logic of Collective Action* (Cambridge, Mass.: Harvard University Press).

Olson, M. Jnr. (1982), *The Rise and Decline of Nations* (New Haven: Yale University Press).

Ouchi, W. G., and Johnson, B. (1978), 'Types of organizational control and their relationship to economic well-being', *Administrative Science Quarterly*, vol. 23, pp. 293–317.

Pahl, R., and Winkler, J. (1974), 'The economic elite: theory and practice', in P. Stanworth and A. Giddens (eds), *Elites and Power in British Society* (Cambridge: Cambridge University Press).

Palmer, R. J., Bignell, B. F., and Levy, J. (1976), *The Education of Graduate Mechanical Engineers* (London: Department of Mechanical Engineering, City University).

Parker, S. R. (1972), *The Future of Work and Leisure* (London: Paladin).

Parker, S. R. (1976), *The Sociology of Leisure* (London: Allen & Unwin).

Parker, S. R., Brown, R. K., Child, J., and Smith, M. A. (1981), *The Sociology of Industry* (London: Allen & Unwin).

Parkin, F. (1974), 'Strategies of social closure in class formation', in F. Parkin (ed.), *The Social Analysis of Class Structure* (London: Tavistock).

Pavitt, K. (ed.) (1979), *Technical Innovation and British Economic Performance* (London: Science Policy Research Unit/Macmillan).

Pavitt, K. (1983), 'Characteristics of innovative activities in British industry', *Omega: The International Journal of Management Science*, vol. 11, pp. 113–38.

Perrucci, R., and Gerstl, J. (1969), *Profession Without Community: Engineers in American Society* (New York: Random House).

Pettigrew, A. M. (1973), *The Politics of Organizational Decision-Making* (London: Tavistock).

Pettigrew, A. M. (1985), *The Awakening Giant: Continuity and Change in Imperial Chemical Industries* (London: Blackwell).

Pinto, A., and Knakal, J. (1973), *El Sistemo Centro-peripheria 20 Anos Despues* (Santiago: Instituto Lantinamericano de Planificacion Economica).

Playfair, L. (Baron) (1852), *Industrial Instruction on the Continent* (London: Royal School of Mines).

Playfair, L. (Baron) (1853), *Industrial Instruction on the Continent* (London: Government School of Mines).

Pollard, S. (1982), *The Wasting of the British Economy: British Economic Policy 1945 to the Present* (London: Croom Helm).

Pollard, S. (1983), *The Development of the British Economy 1914–1980* (3rd edn) (London: Edward Arnold).

Poole, M., and Mansfield, R. (1980), *Managerial Roles in Industrial Relations: Towards a Definitive Survey of Research and Formulation of Models* (Aldershot: Gower).

Poole, M., Mansfield, R., Blyton, P., and Frost, P. (1981), *Managers in Focus* (Aldershot: Gower).

Prabhu, V. B., and Russell, J. W. (1978), 'Survey of production management in north-east manufacturing industry: summary of main findings' (Newcastle: Newcastle-upon-Tyne Polytechnic).

Prabhu, V. B., and Russell, J. (1982), 'The truth about production', *Management Today*, June, pp. 82–160.

Prais, S. J. (1981a), *Productivity and Industrial Structure* (Cambridge: Cambridge University Press).

Prais, S. J. (1981b), 'Vocational qualifications of the labour force in Britain and Germany', *National Institute Economic Review*, no. 98, pp. 47-59.

Prais, S. J., and Wagner, K. (1983), 'Some practical aspects of human capital investment: training standards in five occupations in Britain and Germany', *National Institute Economic Review*, no. 105, August, pp. 46–65.

Prandy, K. (1965), *Professional Employees: A Study of Scientists and Engineers* (London: Faber).

Prandy, K., Stewart, A., and Blackburn, R. M. (1980), *Social Stratification and Occupations* (London: Macmillan).

Prandy, K., Stewart, A., and Blackburn, R. M. (1983), *White Collar Unionism* (London: Macmillan).

Prest, A. R., and Coppock, D. J. (1984), *The UK Economy: A Manual of Applied Economics* (10th edn) (London: Weidenfeld and Nicolson).

Price, D. J. de Solla (1963), *Little Science – Big Science* (New York: Columbia University Press).

Price, D. J. de Solla (1965), 'Is technology historically independent of science? A study in statistical historiography', *Technology and Culture*, vol. 6, pp. 1–35.

Pym, D. (1969), 'Education and the employment opportunities of engineers', *British Journal of Industrial Relations*, vol. 7, pp. 42–51.

Pym, D. (1975), 'The demise of management and the ritual of employment', *Human Relations*, vol. 28, pp. 675–98.

Race Relations Act 1976, *Elizabeth II, 1976*, ch. 74 (London: HMSO).

Rae, J. B., 'Engineers are people' (1975), *Technology and Culture*, vol. 16, pp. 404–18.

Reader, W. J. (1966), *Professional Men: The Rise of the Professional Classes in Nineteenth-Century England* (London: Weidenfeld & Nicolson).

Resale Prices Act 1976, *Elizabeth II, 1976*, ch. 53 (London: HMSO).

Restrictive Trade Practices Act 1976; *Elizabeth II, 1976*, ch. 34 (London: HMSO).

Ritti, R. R. (1971), *The Engineer in the Industrial Corporation* (New York: Columbia University Press).

Rock, P. (1979), *The Making of Symbolic Interactionism* (London: Macmillan).

Rolt, L. T. C. (1970), *Victorian Engineering* (Harmondsworth: Penguin).

Rose, A. M. (ed.) (1962), *Human Behaviour and Social Process: An Interactionist Approach* (London: Routledge & Kegan Paul).

Rose, M. (1975), *Industrial Behaviour: Theoretical Developments Since Taylor* (Harmondsworth: Allen Lane).

Roslender, R. (1983), 'The Engineers' and Managers' Association', *Industrial Relations Journal*, vol. 14, 2, pp. 41–51.

Roszak, T. (1970), *The Making of a Counter-Culture: Reflections on the Technocratic Society and its Youthful Opposition* (London: Faber).

Rothschild, N. (1971), 'The organization and management of government research and development', in *A Framework for Government Research and Development: A Green Paper*, Cmnd 4814 (London: HMSO).

Rothwell, R. (1981), 'Technology, structural change and manufacturing employment', *Omega: An International Journal of Management Science*, vol. 9 (3), pp. 229–45.

Routh, G. (1980), *Occupation and Pay in Great Britain, 1906–79* (London: Macmillan).

Roy, A. D. (1982), 'Labour productivity in 1980: an international comparison', *National Institute Economic Review*, vol. 101 (August), pp. 26–37.

Roy, D. (1952), 'Quota restriction and goldbricking in a machine shop', *American Journal of Sociology*, vol. 57, pp. 427–42.

Roy, D. (1953), 'Work satisfaction and social reward in quota achievement: an analysis of piecework incentive', *American Sociological Review*, vol. 18, pp. 507–14.

Roy, D. (1969), 'Making out: a workers counter-system of control of work situation and relationships', in T. Burns (ed.), *Industrial Man* (Harmondsworth: Penguin) (from 'Efficiency and the fix: informal intergroup relations in a piecework machine shop', *American Journal of Sociology*, vol. 60 (1955), pp. 255–66).

Royal Commission on Technical Education (1882, 1884), *Report of The Royal Commission on Technical Instruction, Parliamentary Papers*, 1882, vol. 27; 1884, vols 29 and 31.

Rush, H. M. F. (1969), 'Behavioural science concepts and management applications', *Studies in Personnel Policies No. 210* (New York: National Industrial Conference Board).

Salmond, J. W., Heuston, R. F. V., and Chambers, R. S. (1981), *Law of Torts* (London: Sweet & Maxwell).

Samuelson Committee (1867–8), *Report of the Endowed Schools (Schools Inquiry) Commission*, Cmnd 3966, vol. 1.

Sargent, J. R. (1978), 'UK Performances in Services', in Blackaby, F. (ed.), *De-industrialisation* (London: Heinemann).

Scott, J. (1979), *Corporations, Classes and Capitalism* (London: Hutchinson).

Scott, J. P. (1982), *The Upper Classes: Property and Privilege in Britain* (London: Macmillan).

Selwyn, N. M. (1982), *Law of Employment* (London: Butterworth).

Sex Discrimination Act 1975, *Elizabeth II, 1975*, ch. 65 (London: HMSO).

Shanks, M. (1971), *The Stagnant Society* (Harmondsworth: Penguin).

Shanks, M. (1978), *What's Wrong with the Modern World?* (London: Bodley Head).

Sherif, P. (1976), *Career Patterns in the Higher Civil Service* (London: HMSO).

Sideri, S. (1972), 'International trade and economic power', in J. Debrandt *et al.* (eds), *Towards a World Economy* (Rotterdam: University Press).

Silverman, D. (1970), *The Theory of Organisations: a Sociological Framework* (London: Heinemann).

Sisson, K. (1983), 'Employers' organizations', in W. Brown (ed.), *Industrial Relations in Britain* (Oxford: Blackwell).

Sklair, L. (1973), *Organised Knowledge: A Sociological View of Science and Technology* (London: Hart-Davis MacGibbon).

Smith, A. (1976), *An Inquiry into the Nature and Causes of the Wealth of Nations*, ed. R. H. Campbell and A. S. Skinner (Oxford: Clarendon).

Smith, A. (1977), 'Education and economic well being', *Engineering, Technology and Society: Proceedings of Section X of the British Association for the Advancement of Science*, Lancaster, 1976 (Birmingham: University of Aston).

Smith, C. S. (1970), 'Art, technology and science', *Technology and Culture*, vol. 11, October, pp. 493–549.

Smith, M., McLoughlin, J., Large, P., and Chapman, R. (1985), *Asia's New Industrial World* (London: Methuen, 1985).

Smithers, A. (1969), 'Occupational aspirations and expectations of engineering students on sandwich courses', *British Journal of Industrial Relations*, vol. 7, p. 414–22.

Snow, C. P. (1966), 'The place of the engineer', *Chartered Mechanical Engineer*, May, pp. 215–18 (12th Graham Clark Lecture).

Snow, Lord (1969), *The Two Cultures and a Second Look: An Expanded Version of the Two Cultures and the Scientific Revolution* (Cambridge: Cambridge University Press).

Sofer, C. (1970), *Men in Mid-Career: A Study of British Managers and Technical Specialists* (Cambridge: Cambridge University Press).

Sorge, A. (1978a), 'Management, technical education and training as a public concern in Britain, France and Germany', Discussion Paper no. 78–33 (Berlin: International Institute of Management).

Sorge, A. (1978b), 'The management tradition – a continental view', in M. Fores and I. Glover (eds), *Manufacturing and Management* (London: HMSO).

Sorge, A. (1978c), 'Technique and culture', Discussion Paper (Berlin: International Institute of Management).

Sorge, A. (1979), 'Engineers in management: a study of the British, German and French traditions', *Journal of General Management*, vol. 5, pp. 46–57.

Sorge, A. (1985), 'Culture's consequences', in P. Lawrence and K. Elliott (eds), *Introducing Management* (Harmondsworth: Penguin).

Sorge, A., and Hartmann, G. (1980), 'Technology and labour markets', Discussion Paper No. 80–39 (Berlin: International Institute of Management).

Sorge, A., Hartmann, G., Warner, M., and Nicholas, I. (1983), *Microprocessors in Manufacturing* (Aldershot: Gower).

Sorge, A. (previously Röder-Sorge), and Nagels, K. (1977), *Industrielle Demokratie in Europa: Mitbestimmung und Kontrolle in der Europaischen-Aktiengesellschaft* (Frankfurt: Rotterdam).

Sorge, A., and Warner, M. (1977), *Variety and Determinants of Factory Organisation in Britain, France and Germany: UK National Report* (Henley and Uxbridge: Administrative Staff College and Brunel University).

Sorge, A., and Warner, M. (1980), 'Manpower training, manufacturing organisation and workplace relations in Great Britain and West Germany', *British Journal of Industrial Relations*, vol. 18 (November), pp. 318–33.

Sorge, A., and Warner, M. (1986), *Comparative Factory Organisation* (Aldershot: Gower).

Spencer, A. (1981), 'What executives don't do', *Management Today* (May), pp. 51–3.

Spencer, A., and McAuley, J. (1980), 'The role of the non-executive director: a *Verstehen* Approach', *Journal of Management Studies*, vol. 17, 1, pp. 82–95.

Stanic, V., and Pym, D. (1968), *Brains Down the Drain*, Anbar Monograph No. 12 (London: Anbar).

Stanworth, P. and Giddens, A. (eds) (1974), *Elites and Power in British Society* (Cambridge: Cambridge University Press).

Stewart, R. (1967), *Managers and Their Jobs* (London: Macmillan).

Storey, J. (1983), *Managerial Prerogative and the Question of Control* (London: Routledge & Kegan Paul).

Storr, A. (1968), *Human Aggression* (Harmondsworth: Penguin).

Swann, N. (1968), *The Flow into Employment of Scientists, Engineers and Technologists*, Cmnd 3760 (The Swann Report) (London: HMSO).

Swinnerton-Dyer, P. (1982), *Report of the Working Party on Post-Graduate Education*, Cmnd 8537 (London: HMSO).

Swords-Isherwood, N. (1979), 'British Management Compared', in K. Pavitt (ed.), *Technical Innovation and British Economic Performance* (London: Science Policy Research Unit, University of Sussex/Macmillan).

Taguiri, R. (1965–6), 'Value orientations and the relationships of managers and scientists', *Administrative Science Quarterly*, vol. 10, pp. 38–51.

Tanzer, M. (1980), *The Race for Resources* (London: Heinemann).

Taylor, F. W. (1964), *Scientific Management* (New York: Harper & Row).

Taylor, R. (1978), *The Fifth Estate: Britain's Unions in the Seventies* (London: Routledge & Kegan Paul).

Taylor, R. (1979), 'Career orientations and intra-occupational choice: a survey of engineering students', *Journal of Occupational Psychology*, vol. 52, pp. 41–52.

Taylor, R. (1982), *Workers and the New Depression* (London: Macmillan).

Terkel, S. (1977), *Working* (Harmondsworth: Penguin).

Thackray, J. (1978), 'America's output problem', *Management Today* (June), pp. 78–81.

Thackray, J. (1981a), 'America's mish-mash managements', *Managements Today* (April), pp. 60–5.

Thackray, J. (1981b), 'America's technology gap', *Management Today* (June), pp. 68–73.

Thickett, L. A. (1971), 'A long look at Honda: with the aim of comparing the operations of the world's biggest manufacturers with those of the UK industry', *Motor Cycle Sport* (October), pp. 386–95.

Thomason, G. F. (1981), *A Textbook of Personnel Management* (London: Institute of Personnel Management).

Thurley, K. (1982), 'The Japanese model: practical reservations and surprising opportunities', *Personnel Management* (February), pp. 36–9.

258   *Engineers in Britain*

Thust, P. (1980), 'The engineer as manager', *European Journal of Engineering Education*, vol. 4, pp. 167–83.

Tinbergen, N. (1951), *The Study of Instinct* (Oxford: Oxford University Press).

Tinbergen, N. (1968), 'On war and peace in animals and men: an ethologist's approach to the biology of aggression', *Science*, vol. 160, pp. 1411–18.

Tobias, S. A. (1968), *Mechanical Engineering* (Cambridge: Advisory Centre for Education).

Torrington, D., and Chapman, J. (1979), *Personnel Management* (Englewood Cliffs, NJ: Prentice-Hall).

Torrington, D., and Weightman, J. (1982), 'Technical atrophy in middle management', *Journal of General Management*, vol. 7 (4), pp. 5–17.

Trade Union and Labour Relations Act 1974, *Elizabeth II, 1974*, ch. 52 (London: HMSO).

Trist, E. L., and Bamforth, K. W. (1951), 'Some social and psychological consequences of the Longwall method of coal-getting', *Human Relations*, vol. 4, pp. 3–38.

Trist, E. L., Higgin, G. W., Murray, H., and Pollock, A. B. (1963), *Organizational Choice* (London: Tavistock).

Tubman, K. A., and Lewis, J. S. (1979), 'Attracting able youths to careers in engineering and technology', *Chartered Mechanical Engineer* (February), pp. 40–2.

Tunstall, J. (1962), *The Fishermen* (London: McGibbon & Kee).

Turner, B. T., and Williams, M. R. (1983), *Management Handbook for Engineers and Technologists* (London: Business Books).

Urry, J. (1973), 'Towards a structural theory of the middle classes', *Acta Sociologica*, vol. 16, pp. 175–87.

Veblen, T. (1921), *The Engineers and the Price System* (New York: Viking).

Veblen, T. (1953), *The Theory of the Leisure Class* (New York: Mentor).

Vittas, D. (1986), 'Bankers' relations with industry: an international survey', *National Westminster Bank Quarterly Review* (February), pp. 2–14.

Vroom, B. (1964), *Work and Motivation* (New York: Wiley).

Vroom, V., and Deci, E. L. (1970), *Management and Motivation* (Harmondsworth: Penguin).

Waddington, I. (1973), 'The role of the hospital in the development of modern medicine: a sociological analysis', *Sociology*, vol. 7, pp. 211–24.

Walsh, K., Hinings, R., Greenwood, R., and Ransom, S. (1981), 'Power in organizations', *Organization Studies*, vol. 2, pp. 131—52.

Watson, H. B. (1976), 'Organizational bases of professional status: a comparative study of the engineering profession', unpublished PhD thesis, University of London.

Watson, T. J. (1977), *The Personnel Managers: A Study in the Sociology of Work and Employment* (London: Routledge & Kegan Paul).

Watson, T. J. (1980), *Sociology, Work and Industry* (London: Routledge & Kegan Paul).

Webb, S., and Webb, B. (1920), *The History of Trade Unionism 1666–1920* (London: Longman).

Weber, M. (1947), *The Theory of Social and Economic Organization*, introduction by T. Parsons (New York: Free Press; London: Collier Macmillan).

Weber, M. (1948), 'Class, status, party', in H. H. Gerth and C. W. Mills (eds), *From Max Weber: Essays in Sociology* (London: Routledge & Kegan Paul, pp. 180–95).

Wedderburn, K. W., Lewis, R., and Clark, J. (1983), *Labour Law and Industrial Relations: Building on Kahn-Freund* (Oxford: Oxford University Press).

Whalley, P. (1984), 'Deskilling engineers? The labor process, labor markets, and labor segmentation', *Social Problems*, vol. 32, 1 (December), pp. 117–32.

Whalley, P. (1986), *The Social Production of Technical Work: The Case of British Engineers* (London: Macmillan).

Whincup, M. (1983), *Modern Employment Law: A Guide to Job Security and Safety* (4th edn) (London: Heinemann).

Whitley, R., Thomas, A., and Marceau, J. (1981), *Masters of Business? Business Schools and Business Graduates in Britain and France* (London: Tavistock).

Wiener, M. (1981), *English Culture and the Decline of the Industrial Spirit 1850–1980* (Cambridge: Cambridge University Press).

Wilby, P. (1985), *The Sunday Times Good Career Guide* (London: Granada).

Wild, R. (1982), *How to Manage: 123 World Experts Analyse the Art of Management* (London: Heinemann).

Wilkinson, R. (1964), *The Prefects: British Leadership and the Public School Tradition* (Oxford: Oxford University Press).

Wotjas, O. (1984), 'Scotland's great debate', *The Times Higher Education Supplement*, 14 December, pp. i–iv.

Wood, S. (1982), *The Degradation of Work? Skill, Deskilling and the Labour Process* (London: Hutchinson).

Woodward, L. (1962), *The Age of Reform 1815–1870* (2nd edn) (Oxford: Clarendon).

Zeitlin, I. (1967), *Marxism: A Re-examination* (London: Van Nostrand).

Zeitlin, M. (1974), 'Corporate ownership and control', *American Journal of Sociology*, vol. 79, no. 5, pp. 1073–1119.

Subject Index

Author Index